FOR GRADE
+SAT PREP

Kids SUMMER ACADEMY

ARGOPREP

7 DAYS A WEEK
8 WEEKS

- Mathematics
- Logic
- English
- SAT Prep
- Reading
- Writing

GRADE 10-11

ArgoPrep is one of the leading providers of supplemental educational products and services. We offer affordable and effective test prep solutions to educators, parents and students. Learning should be fun and easy! To access more resources visit us at www.argoprep.com.

Our goal is to make your life easier, so let us know how we can help you by e-mailing us at: info@argoprep.com.

- ArgoPrep is a recipient of the prestigious **Mom's Choice Award**.
- ArgoPrep also received the 2019 **Seal of Approval** from Homeschool.com for our award-winning workbooks.
- ArgoPrep was awarded the 2019 **National Parenting Products Award**, **Gold Medal Parent's Choice Award** and **the Tillywig Brain Child Award.**

ISBN: 978-1962936262
Published by Argo Brothers.

TABLE OF CONTENTS

TABLE OF CONTENTS

HOW TO USE THE BOOK

Welcome to **Kids Summer Academy** by ArgoPrep.

This workbook is designed to prepare students over the summer to get ready for **Grade 11**.

The curriculum has been divided into **eight weeks** so students can complete this entire workbook over the summer.

Our workbook has been carefully designed and **crafted by licensed teachers** to give students an incredible learning experience.

Students start off the week with English activities followed by Math practice. Throughout the week, students have several fitness activities to complete. Making sure students stay active is just as important as practicing mathematics.

We introduce basic fitness activities that any student can complete. On the last day of each week, students will work on a logic games.

KIDS SUMMER ACADEMY SERIES

ArgoPrep's **Kids Summer Academy** series helps prevent summer learning loss and gets students ready for their new school year by reinforcing core foundations in math, english and science. Our workbooks also introduce new concepts so students can get a head start and be on top of their game for the new school year!

CAPTAIN BRAVERY

WATER FIRE

MYSTICAL NINJA

GREEN POISON

FIRESTORM WARRIOR

RAPID NINJA

CAPTAIN ARGO

THUNDER WARRIOR

ADRASTOS THE SUPER WARRIOR

DANCE HERO

GREEN DRAGON WARRIOR

WEEK 1

GRADE 10-11

This week, you will learn about the three core rhetorical techniques - ethos, pathos, and logos - used by authors to persuade or inform an audience. For math, you will review combining like terms, solving algebraic equations, absolute value equations, and scientific notation.

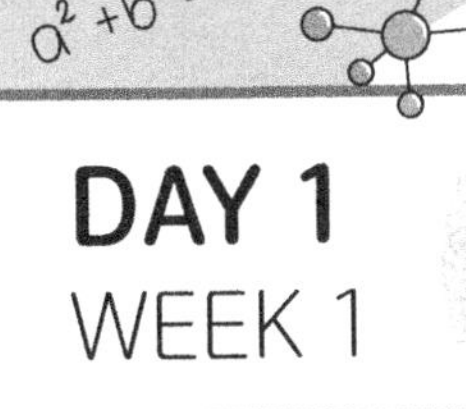

DAY 1
WEEK 1

ELA
RHETORICAL TECHNIQUES

Welcome to our Summer Academy Series to get ready for 11th grade! Today, we'll learn about rhetorical techniques.

Rhetorical techniques are strategies used by authors to persuade or inform an audience, typically involving appeals to emotion, ethics, or logic. These techniques can make your arguments more convincing and help you communicate your ideas more clearly. As an 11th grader, understanding rhetorical techniques is critical as it will allow you to build your persuasive writing and speaking skills.

The Three Core Rhetorical Techniques:

1. Ethos (Ethical Appeal):

- Ethos appeals to the audience's perception of the speaker's credibility and character. An author or speaker conveys this persona by showing knowledge of the subject, fairness, and goodwill.
- Effect: Enhances trust and authority; if the audience believes the speaker is reliable and ethical, they are more likely to be persuaded.
- Example: "As a doctor with over 30 years of experience specializing in immunology, I strongly recommend this vaccine for its efficacy and safety." The speaker establishes their credentials and long-standing expertise in the field, which builds trust with the audience.

2. Pathos (Emotional Appeal):

- Pathos appeals to the audience's emotions, trying to trigger feelings of pity, anger, joy, or other strong emotions to persuade.
- Effect: Helps connect the audience emotionally with the subject; powerful emotional responses can override logic.
- Example: "Imagine a child, born into poverty, who walks five miles each day to school under the scorching sun and still manages to smile. Your donation can change that child's life." The vivid description evokes empathy and compassion, motivating the audience to act.

3. Logos (Logical Appeal):

- Logos appeals to logic and reason. It involves presenting facts, statistics, logical arguments, and rationality to persuade the audience.
- Effect: Strengthens argument by utilizing factual evidence and logical reasoning, aiming to appeal to the audience's rational thinking.
- Example: "Research shows that students who participate in extracurricular activities score 10% higher on standardized tests compared to those who do not." By citing specific statistics, the argument becomes more compelling and grounded in logic.

Directions: Read the passage below and then answer the questions that follow.

The Importance of Adopting a Pet from a Shelter

As a veterinarian with over 15 years of experience, I have witnessed firsthand the countless benefits of adopting a pet from a local animal shelter. When you choose to adopt, you are not only providing a loving home to an animal in need but also supporting a cause that helps reduce pet overpopulation and homelessness.

Imagine the fear and loneliness that a shelter animal experiences, being confined to a small space without the comfort of a familiar family. These animals often come from difficult backgrounds, having been abandoned, surrendered, or rescued from abuse. By adopting a pet, you have the power to transform their lives, offering them the love, security, and companionship they desperately crave.

Moreover, adopting from a shelter is a financially responsible decision. Shelter animals are typically already spayed or neutered, vaccinated, and microchipped, which can save you hundreds of dollars in veterinary expenses. Additionally, most shelters charge a modest adoption fee, which is significantly lower than the cost of purchasing a pet from a breeder or pet store.

Furthermore, studies have shown that owning a pet can provide numerous physical and mental health benefits. Pet owners tend to have lower blood pressure, reduced stress levels, and increased opportunities for exercise and socialization. By adopting a pet, you are not only improving their life but also enhancing your own well-being.

In conclusion, adopting a pet from a shelter is a compassionate, economically sound, and mutually beneficial decision. By opening your heart and home to a shelter animal, you are making a difference in their life and enriching your own. Visit your local animal shelter today and discover the joy of adoption.

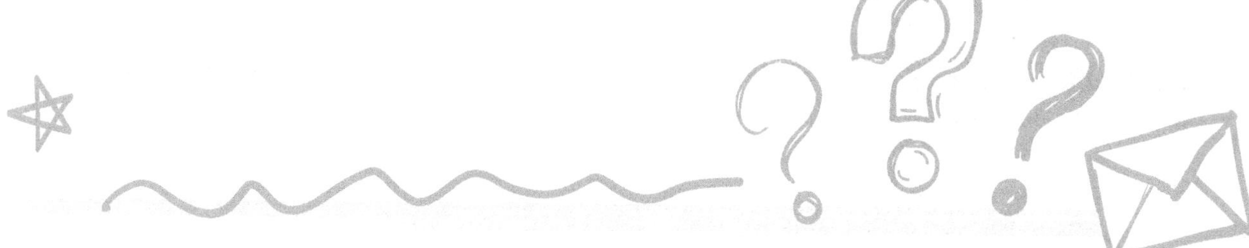

DAY 1
WEEK 1

ELA
RHETORICAL TECHNIQUES

Question 1:

Identify an example of ethos used in the passage and explain how you know the sentence identified uses ethos to persuade the reader.

Question 2:

Identify an example of pathos used in the passage and explain how you know the sentence identified uses pathos to persuade the reader.

Question 3:

Identify an example of logos used in the passage and explain how you know the sentence identified uses logos to persuade the reader.

Let's get some fitness in! Go to page 169 to try some fitness activities.

DAY 2
WEEK 1

ELA
RHETORICAL TECHNIQUES

Directions: Read each excerpt carefully. Choose the best answer from the provided options that identifies the primary rhetorical appeal (ethos, pathos, or logos) being utilized.

Question 1:

An advertisement for life insurance shows a happy family playing in the park and emphasizes, "Because you care about your family's future."

Which rhetorical appeal is primarily being used?

A. Ethos
B. Pathos
C. Logos

Question 2:

A politician begins a speech with the following words: "As a veteran and someone who has served this country for over 20 years, I understand the importance of national security better than anyone."

Which rhetorical appeal is primarily being used?

A. Ethos
B. Pathos
C. Logos

Question 3:

A lawyer presents the following argument: "According to the data collected from over 100 cases in the past year, 95% of individuals who faced this issue received a favorable ruling by taking this specific legal action."

Which rhetorical appeal is primarily being used?

A. Ethos
B. Pathos
C. Logos

Question 4:

An environmental organization's pamphlet states, "Imagine a world where your children can breathe clean air and drink pure water. By joining our cause, you can make this a reality."

Which rhetorical appeal is primarily being used?

A. Ethos
B. Pathos
C. Logos

Question 5:

Which rhetorical appeal is most evident in a charity's plea that shows images of animals in distress along with a call to action to donate today?

A. Ethos
B. Pathos
C. Logos

Question 6:

Select the scenario where ethos is the dominant rhetorical appeal.

A. A lawyer presents eye-witness testimonies and forensic evidence during a trial.
B. A doctor discusses the benefits of a new medical procedure, citing her 20 years of surgical experience.
C. A motivational speaker tells his personal story of overcoming adversity to inspire his audience.
D. A nonprofit organization provides statistical evidence showing the effectiveness of its programs.

Question 7:

Which scenario primarily utilizes ethos to persuade the audience?

A. A novelist writes a blog post explaining the research process behind her historical fiction novel, emphasizing her thorough investigation and collaboration with historians.

B. A tech CEO discusses future trends in technology during a keynote, citing predictions based on current data analysis and market trends.

C. A finance expert gives a webinar on investment strategies, detailing his 30 years of experience in managing successful portfolios.

D. A marketing director uses customer testimonials and satisfaction ratings to promote a new product.

Question 8:

Write a short persuasive essay where **logos** is the primary rhetorical strategy.

Topic: The benefits of learning a second language. You can use the Internet to help you. Your essay should include at least three logical arguments supported by data, statistics, or well-reasoned arguments.

ELA

RHETORICAL TECHNIQUES

Let's get some fitness in! Go to page 169 to try some fitness activities.

DAY 3 WEEK 1

MATH
COMBINING LIKE TERMS

OVERVIEW:

Combining like terms is a fundamental algebraic technique used to simplify expressions and solve equations more efficiently. It involves consolidating terms that have the same variable raised to the same power into a single term, which makes algebraic manipulations and the solution process more straightforward.

Like terms are terms within an algebraic expression that have the same variable components and the same exponents.

For example, in the expression $3x + 4x^2 - 2x$, the terms $3x$ and $-2x$ are like terms because they both contain the variable x raised to the first power. Since $3x$ and $-2x$ are **like terms** we can simplify by combining $3x - 2x$ which gives us $1x$ or just x.

Take a look at this following expression and simplify it by combining like terms.

$$4y - 2y + 3x - x + 6$$

We can see that $4y$ and $-2y$ are like terms. $3x$ and $-x$ are also like terms.

$$4y - 2y = 2y$$

$$3x - x = 2x$$

We are left with the simplified expression $2y + 2x + 6$.

Look at the guided practice problems below.

Simplify the expression.

Example 1: $5b - 8 - 5b + 7$

The like terms are $5b$ and $-5b$, which just cancel each other out. The other like terms are -8 and $+7$, which is $-8 + 7 = -1$.

This expression simplified is -1.

Example 2: $-46p - 44(2 - 3p)$

We need to first distribute before combining like terms. $-44(2 - 3p)$ distributed is $-88 + 132p$. We can re-write the original expression to $-46p - 88 + 132p$. Now, let's combine like terms which are $-46p + 132p$, giving us $86p - 88$.

The answer is $86p - 88$.

Your turn to practice simplifying expressions!

DAY 3
WEEK 1

MATH
COMBINING LIKE TERMS

Simplify each expression. Do NOT use a calculator.

1. $1 + 67a - 65$

A. $1 - a$
B. $-64 + 67a$
C. $-103a$
D. $-66a$

2. $-29n - 79n$

A. $-98n$
B. $65n + 20$
C. $107n + 20$
D. $-108n$

3. $66n - 94 + 66n$

A. $132n - 94$
B. $-65n$
C. $-157 - 5n$
D. $158n - 94$

4. $46x + 35x(-50x - 54)$

A. $-40x + 1620$
B. $-184x^2 - 6764x$
C. $1321 - 1848x$
D. $-1844x - 1750x^2$

5. $97 - 32(-4 - 43a)$

A. $3391a - 46a^2$
B. $225 + 1376a$
C. $4740a + 4748a^2$
D. $422a - 5760a^2$

6. $23(-18 - 67n) + 56$

A. $-358 - 1641n$
B. $-358 - 1541n$
C. $3970 - 98n$
D. $-358 - 1667n$

7. $-44(k + 48) + 18(46 + 10k)$

A. $270k + 551k^2 - 1450$
B. $271k + 551k^2 - 1450$
C. $287k + 551k^2 - 1450$
D. $136k - 1284$

8. $-25x(1 + 36x) + 13(3x + 15)$

A. $14x - 900x^2 + 195$
B. $956 + 781x$
C. $-16x^2 + 68x + 102$
D. $-16x^2 + 115x + 102$

9. $20k(k + 12) - 32(10k + 41)$

A. $20k^2 - 80k - 1312$
B. $61k + 335$
C. $15k + 335$
D. $93k + 335$

10. $-40n(42 - 6n) + 5(20n - 44)$

A. $-1580n + 240n^2 - 220$
B. $38 + 1328n - 1250n^2$
C. $448n - 1715n^2 - 735$
D. $38 + 1281n - 1250n^2$

DAY 4
WEEK 1

MATH
ALGEBRAIC EQUATIONS

OVERVIEW:

An algebraic equation consists of two expressions set equal to each other, involving variables (usually represented by letters like x or y) and constants (numerical values). The simplest form of an algebraic equation is the linear equation in one variable, such as:

$ax + b = c$ where a, b, and c are constants, and x is the variable.

Basic Principles for Solving Algebraic Equations

1. Balance Principle: The balance principle states that you can perform the same operation (addition, subtraction, multiplication, division) on both sides of the equation without changing the equality. In other words: **Whatever you do to one side of an equation, you must do to the other side.**

2. Inverse Operations: Use inverse operations to isolate the variable. For addition, use subtraction; for multiplication, use division, and vice versa.

3. Order of Operations: Remember to follow the order of operations (PEMDAS) when solving equations that involve multiple steps.

Look at the guided practice problems below.

Example 1: $34 + 5m = -7(m - 10)$

We need to first distribute the right side of the equation. $-7(m - 10)$ distributed gives us $-7m + 70$ on the right side of the equation.

We can re-write the equation to $34 + 5m = -7m + 70$.

We need to isolate the variable, so let's first subtract 34 from both sides of the equation to give us $5m = -7m + 36$. Next, we can add 7m to both sides to get $12m = 36$. Finally, we can divide both sides by 12, successfully isolating the variable, giving us $m = 3$.

Example 2: $-\frac{37}{10}x + 2x = -\frac{119}{50}$

In this equation, we can see on the left side we have like terms that can be combined together. $-\frac{37}{10}x + 2x$ ← We can re-write this as $-\frac{37}{10}x + \frac{2x}{1}$ (Remember we must have a common denominator when adding fractions. The second fraction needs to be multiplifed by 10 on the numerator and denominator.)

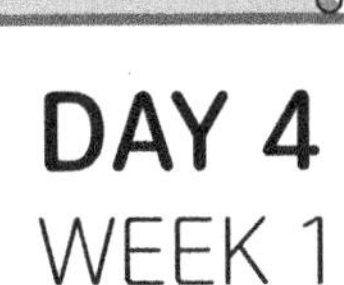

$$-\frac{37}{10}x+\frac{2x(10)}{1(10)}=-\frac{37}{10}x+\frac{20x}{10}=-\frac{17}{10}x$$

We can re-write our original equation to $-\frac{17}{10}x=-\frac{119}{50}$

Now, we can solve for x by dividing both sides by $-\frac{17}{10}$.

$$x=-\frac{119}{50}\div-\frac{17}{10}$$

$$x=-\frac{119}{50}\times-\frac{10}{17}$$

I can re-write this as

$x=\frac{119\times5\times2}{10\times5\times17}$ ← Notice we have negative times negative which is just positive. To help us simplify, the number 10 on the numerator was rewritten as 5 × 2 and the number 50 on the denominator was rewritten to 10 × 5.

Cancel out the common factors and simplify.

$$x=\frac{7}{5}$$

Our answer is $x=\frac{7}{5}$. You can always verify the solution by substituting the value back into the original equation and check if both sides are equal.

Your turn to practice solving algebraic equations!

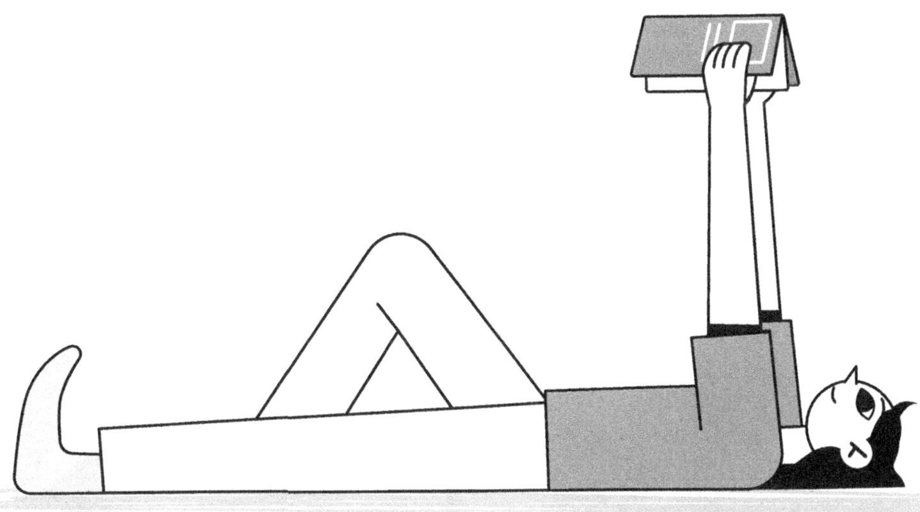

DAY 4
WEEK 1

MATH
ALGEBRAIC EQUATIONS

Solve each equation. Do NOT use a calculator. Use a separate piece of paper to show all your work and double-check your work by substituting the value back into the original equation.

1. $k + 8 + 6 = 7(4k - 4) + 3(5 - 8k)$

2. $-6x + x = -8(x + 3) + 6(6x - 7)$

3. $n - 4 + 1 - 2n = -4(5 - 2n) - 4(-4 + 2n)$

4. $4(6b + 3) + 6(2 - 4b) = -6b - 6b$

5. $-2(1 + 2x) + 6x = -2x - 6(x + 7)$

6. $-2x + \frac{5}{2} + \frac{1}{2}x = 8$

7. $p + \frac{2}{3} - \frac{3}{2} = -\frac{7}{3}$

8. $-\frac{3}{2}p - \frac{5}{2}p = 6$

9. $-2r - \frac{2}{3} + \frac{1}{2} = \frac{19}{6}$

10. $2b - \frac{1}{3} + \frac{1}{2}b = \frac{11}{12}$

11. $-3\left(3b - \frac{3}{2}\right) = \frac{69}{2}$

12. $-1 + 5\left(\frac{10}{3}x + \frac{8}{3}\right) = \frac{661}{9}$

13. $-\frac{14}{5}\left(-\frac{26}{5}x - \frac{17}{6}\right) = -\frac{973}{15}$

14. $-5\left(2x - \frac{7}{4}\right) = \frac{845}{12}$

15. $\frac{226}{3} = -4\left(5m - \frac{4}{3}\right)$

Let's get some fitness in! Go to page 169 to try some fitness activities.

DAY 5
WEEK 1

MATH
ABSOLUTE VALUE EQUATIONS

OVERVIEW:

Absolute value equations are mathematical expressions that involve the absolute value of a number or an expression. The absolute value of a number is the **distance** of that number from zero on the number line.

For example, the absolute value of -5 is 5, and the absolute value of 5 is also 5. The symbol for absolute value is two vertical bars, like this: | |.

When solving absolute value equations, the goal is to **isolate** the absolute value expression on one side of the equation, and then split the equation into two separate cases: one where the expression inside the absolute value bars is positive, and one where it is negative. These two cases must be solved separately, as the solution to the absolute value equation is the set of all values that make the equation true.

The process of solving absolute value equations can be broken down into several steps:

1. Remove the absolute value bars: To remove the absolute value bars, we must split the equation into two separate cases: one where the expression inside the absolute value bars is positive, and one where it is negative. For example, if the equation is $|x + 3| = 5$, we can split it into two separate equations: $x + 3 = 5$ and $x + 3 = -5$.

2. Solve each case separately: After splitting the equation into two separate cases, we must solve each case separately. In the first case, $x + 3 = 5$, we can subtract 3 from both sides of the equation to get $x = 2$. In the second case, $x + 3 = -5$, we can again subtract 3 from both sides to get $x = -8$.

3. Check the solutions: Once we have solved both cases, we must check our solutions to ensure that they satisfy the original equation. To do this, we plug our solutions back into the original equation and see if both sides of the equation are equal. If they are equal, our solutions are correct.

4. Write the final solution: The final solution to an absolute value equation is the set of all values that make the equation true. In our example, the final solution is $x = 2$ or $x = -8$.

DAY 5 WEEK 1

MATH
ABSOLUTE VALUE EQUATIONS

Look at the guided practice problems below.

Example 1:

$$|x - 3| = 3$$

$$x - 3 = 3 \qquad x - 3 = -3$$
$$+3 \quad +3 \qquad +3 \quad +3$$
$$x = 6 \qquad x = 0$$

Step 1: We must split the equation into two separate cases: one where the expression inside the absolute value bars is positive, and one where it is negative.

$x - 3 = 3$ and $x - 3 = -3$.

Step 2: Solve and you see that $x = 6$ and $x = 0$.

Example 2:

$$|37n + 50| = 1234$$

$$37n + 50 = 1234 \qquad 37n + 50 = -1234$$
$$-50 \quad -50 \qquad -50 \quad -50$$
$$\frac{37n}{37} = \frac{1184}{37} \qquad \frac{37n}{37} = \frac{-1284}{37}$$
$$n = 32 \qquad n = -\frac{1284}{37}$$

Step 1: We must split the equation into two separate cases: one where the expression inside the absolute value bars is positive, and one where it is negative.

Step 2: Solve and you see that $n = 32$ and $n = -\frac{1284}{37}$.

Your turn to practice solving absolute value equations!

DAY 5
WEEK 1

MATH
ABSOLUTE VALUE EQUATIONS

Solve each equation.

1. $|54k + 6| = 372$

2. $|52 + 33r| = 1570$

3. $|39 + 50v| = 1461$

4. $|54k + 6| = 2706$

5. $|-60n - 21| = 459$

6. $|50 - 56v| = 3018$

7. $|-29n - 16| = 1089$

8. $|12 + 42a| = 1776$

9. $|-47x - 58| = 2292$

10. $|10 - 49x| = 88$

11. $\left|-\frac{10}{3}n + \frac{1}{2}\right| = \frac{11}{2}$

12. $\left|-\frac{3}{2}x + \frac{1}{2}\right| = \frac{7}{2}$

13. $\left|-\frac{11}{3}x - \frac{7}{3}\right| = \frac{65}{9}$

14. $\left|\frac{3}{2}r + \frac{7}{3}\right| = \frac{16}{3}$

15. $\left|b - \frac{1}{2}\right| = \frac{19}{6}$

Let's get some fitness in! Go to page 169 to try some fitness activities.

FITNESS TIME

DAY 6
WEEK 1

MATH
SCIENTIFIC NOTATION

Scientific notation is a way of expressing numbers that are either very large or very small in a more convenient and compact form. In scientific notation, a number is written as a **mantissa** (a number between 1 and 10) multiplied by a power of 10. The power of 10 represents the number of zeroes in the original number.

For example, the number 0.000001 (1 millionth) can be written in scientific notation as 1×10^{-6}. The number 100,000,000 can be written as 1×10^{8}.

Here's a few examples of numbers that are written in scientific notation:

- $0.000005 = 5 \times 10^{-6}$
- $7500 = 7.5 \times 10^{3}$
- $0.0003 = 3 \times 10^{-4}$
- $50000000 = 5 \times 10^{7}$
- $0.00000125 = 1.25 \times 10^{-6}$

It is important to note that when performing mathematical operations with numbers in scientific notation, the exponent **must** be the same for both numbers before the operation can take place. For example, to add or subtract two numbers, their exponents must be the same. To multiply or divide two numbers, you simply add or subtract the exponents, respectively.

For example:

$$(5 \times 10^{8}) + (2 \times 10^{8}) = (5 + 2) \times 10^{8} = 7 \times 10^{8}$$

$$(5 \times 10^{8}) \times (2 \times 10^{4}) = (5 \times 2) \times (10^{8} \times 10^{4}) = 10 \times 10^{12} = 10 \times 10^{12}$$

Here are a few rules and examples of how to simplify expressions in scientific notation:

- **Adding or subtracting two numbers:** If the exponents are the same, you can add or subtract the mantissas and keep the exponent the same.

  For example, $(5 \times 10^{8}) + (2 \times 10^{8}) = (5 + 2) \times 10^{8} = 7 \times 10^{8}$.

- **Multiplying two numbers:** To multiply two numbers in scientific notation, you multiply the mantissas and add the exponents.

  For example, $(5 \times 10^{8}) \times (2 \times 10^{4}) = (5 \times 2) \times (10^{8} \times 10^{4}) = 10 \times 10^{12}$.

- **Dividing two numbers:** To divide two numbers in scientific notation, you divide the mantissas and subtract the exponents.

DAY 6 WEEK 1

MATH
SCIENTIFIC NOTATION

For example, $\dfrac{(5 \times 10^8)}{(2 \times 10^4)} = \left(\dfrac{5}{2}\right) \times \left(\dfrac{10^8}{10^4}\right) = 2.5 \times 10^4$

- **Raising a number to a power:** To raise a number in scientific notation to a power, you raise the mantissa to that power and multiply the exponent by the power.

  For example, $(3 \times 10^4)^2 = (3^2) \times (10^4)^2 = 9 \times 10^8$

Look at the guided practice problems below.

Example 1:

$(3.13 \times 10^5)\ (4 \times 10^4)$

Notice that we are multiplying two numbers, so we know we need to multiply the mantissas and add the exponents. First, we will multiply 3.13×4 which gives us 12.52.

Next, we add the exponents $(5 + 4) = 9$. So we have 12.52×10^9. **Please note:** this is not the right answer. Although this is correct, it is not written in the proper scientific notation. There must be a decimal point immediately after the first number, so instead of 12.52 we need to get the number to 1.252. To do this, it's simple! To get from 1.252 to 12.52 we need to multiply by 10, meaning we are just adding 1 to the exponent.

So instead of 12.52×10^9 we will re-write this to 1.252×10^{10}, which is the answer.

Example 2:

$$\frac{(3.33 \times 10^4)}{(5.1 \times 10^{-6})}$$

Explanation: Notice that we are dividing two numbers, so we know we need to divide the mantissas and subtract the exponents. For these questions, feel free to use a calculator since you will be dividing numbers with decimals.

Step 1: Divide the mantissas $\dfrac{3.33}{5.1} = 0.6529$

Step 2: Subtract exponents $4 - (-6) = 10$

DAY 6
WEEK 1

MATH
SCIENTIFIC NOTATION

We currently have 0.6529×10^{10}. However, this is **not** written in the proper scientific notation. The mantissa cannot be 0. The first number must be 6 followed by a decimal point, so we are multiplying the mantissa by 10. Since we are doing this, we need to subtract the exponent by 1, making it 6.529×10^{9}.

Example 3:

$(2.3 \times 10^5)^3$

Explanation: We are raising to a power, so we need to follow the rule that we discussed above for raising a number to a power. To raise a number in scientific notation to a power, you raise the mantissa to that power and multiply the exponent by the power.

Step 1: $2.3^3 = 12.167$

Step 2: The exponent is 5 and we are raising to the power of 3. We need to multiply these two numbers. $5 \times 3 = 15$.

We have 12.167×10^{15}, however, this is **not** written in the proper scientific notation. We need to rewrite it to 1.2167×10^{16}.

Your turn to practice simplifying numbers dealing with scientific notation!

DAY 6
WEEK 1

MATH
SCIENTIFIC NOTATION

Simplify and keep the answer in scientific notation. Circle the correct answer choice.

1. $(2.2 \times 10^{-3})(1.3 \times 10^{3})$

A. 2.86×10^{-1}
B. 1.692×10^{-6}
C. 2.86×10^{1}
D. 2.86×10^{0}

2. $(3 \times 10^{1})(5.6 \times 10^{5})$

A. 5.357×10^{-4}
B. 5.357×10^{-5}
C. 1.68×10^{7}
D. 0.168×10^{7}

3. $(6.1 \times 10^{-4})(6.6 \times 10^{6})$

A. 4.026×10^{3}
B. 4.026×10^{-3}
C. 9.242×10^{-11}
D. 4.026×10^{-4}

4. $(2 \times 10^{-2})(4.5 \times 10^{1})$

A. 4.444×10^{4}
B. 4.444×10^{-4}
C. 9×10^{-1}
D. 90×10^{-1}

5. $(2.5 \times 10^{6})(6.1 \times 10^{4})$

A. 1.525×10^{-11}
B. 1.525×10^{11}
C. 4.098×10^{1}
D. 0.4098×10^{1}

6. $\dfrac{9.6 \times 10^{-5}}{6.4 \times 10^{-3}}$

A. 150×10^{-2}
B. 15×10^{-2}
C. 1.5×10^{-2}
D. 6.144×10^{-7}

7. $\dfrac{1.84 \times 10^{-4}}{2.8 \times 10^{-3}}$

A. 0.6571×10^{2}
B. 0.6571×10^{3}
C. 0.6571×10^{-2}
D. 6.571×10^{-2}

8. $\dfrac{(6.3 \times 10^{1})}{(7.17 \times 10^{-4})}$

A. 878.7×10^{4}
B. 878.7×10^{-4}
C. 87.87×10^{4}
D. 8.787×10^{4}

9. $\dfrac{(6.95 \times 10^{-4})}{(3.55 \times 10^{-6})}$

A. 1.958×10^{2}
B. 19.58×10^{-1}
C. 1.958×10^{-1}
D. 19.58×10^{0}

DAY 6
WEEK 1

MATH
SCIENTIFIC NOTATION

10. $\frac{(6 \times 10^3)}{(1.3 \times 10^0)}$

A. 78×10^3
B. 4.615×10^2
C. 4.615×10^3
D. 7.8×10^3

11. $(7.2 \times 10^{-4})^3$

A. 37.32×10^8
B. 3.732×10^{-11}
C. 3.732×10^8
D. 3.732×10^{-10}

12. $(4 \times 10^3)^4$

A. 2.56×10^{-9}
B. 2.56×10^{-10}
C. 2.56×10^{14}
D. 2.56×10^{-14}

13. $(5.9 \times 10^6)^2$

A. 3.481×10^{13}
B. 5.9×10^3
C. 3.481×10^{-3}
D. 3.481×10^3

14. $(4.4 \times 10^3)^6$

A. 72.56×10^{18}
B. 7.256×10^{18}
C. 7.256×10^{21}
D. 4.4×10^{18}

15. $(6.79 \times 10^3)^2$

A. 46.1×10^8
B. 4.61×10^8
C. 4.61×10^7
D. 4.61×10^3

Notes:

Let's get some fitness in! Go to page 169 to try some fitness activities.

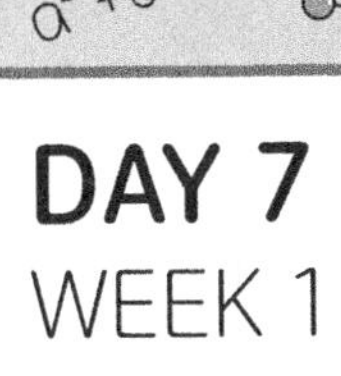

DAY 7
WEEK 1

Logic Games
INTERESTING

1. What number should come next in the series?

1, 4, 9, 16, 25, ?

· o o o ·

2. In a class of **40** students, **18** play basketball, **17** play soccer, and **15** play both sports. How many students in the class do not play either basketball or soccer?

3. If all Zips are Zaps, and no Zap is a Zop, then which of the following must be true?

A. No Zips are Zops.
B. Some Zips are Zops.
C. All Zops are Zips.
D. Some Zops are not Zaps.
E. None of the above.

· o o o ·

4. A man buys a horse for **$60**, sells it for **$70**, buys it back for **$80**, and sells it finally for **$90**. How much has he made?

5. If you're running a race and you pass the person in second place, what place are you in?

WEEK 2

GRADE 10-11

This week, you will practice identifying sentences with a grammatical error and learn about modifiers and clause structures. For math, you will review solving systems of linear equations using either the substitution or elimination method. You will also review the compound interest formula and apply it to word problems.

DAY 1 WEEK 2

ELA GRAMMAR

Directions: Read the paragraph provided below. Each paragraph contains one sentence with a grammatical error. Choose the sentence that has the error from the multiple-choice options.

Question 1:

Despite its modest size, the town's museum offers an array of artifacts that rivals larger institutions. Many visitors are surprised by the breadth and depth of the exhibits, which spans several centuries and various cultures. The museum staff regularly updates the displays to ensure that regular visitors always have something new to experience. Additionally, guided tours are offered every Saturday to enrich visitors' understanding of the historical contexts.

A. Sentence 1

B. Sentence 2

C. Sentence 3

D. Sentence 4

Question 2:

The debate team has won yet another trophy, marking their third consecutive victory this year. In addition to their success, the team has shown incredible dedication and hardwork. Each member practices their speeches rigorously, often staying late after school to refine their delivery and arguments. Their coach, Mr. Thompson, has been a pivotal figure in fostering this culture of excellence and commitment.

A. Sentence 1

B. Sentence 2

C. Sentence 3

D. Sentence 4

Question 3:

When studying the effects of climate change, scientists consider a variety of factors. Changes in temperature, precipitation patterns, and extreme weather events are all taken into account. These changes can affects ecosystems in numerous ways, disrupting food chains and altering habitat conditions. It is crucial for ongoing research to monitor these shifts to better predict future impacts.

A. Sentence 1

B. Sentence 2

C. Sentence 3

D. Sentence 4

Question 4:

The author's latest novel, set in the early 20th century, vividly depicts the struggles and triumphs of its protagonist. As the narrative unfolds, readers are drawn into a world where societal expectations often conflicted with personal desires. His use of symbolic imagery and complex characters enhances the depth of the story. Overall, the book has been received positively, although some critics argues that the pacing is too slow.

A. Sentence 1
B. Sentence 2
C. Sentence 3
D. Sentence 4

Question 5:

The city council recently passed a new ordinance aimed at reducing public disturbances. One of the provisions require that all events with over 100 attendees obtain a special permit. The goal is to ensure that events are properly managed and do not interfere with the daily activities of local residents. This decision has sparked a debate among citizens, many of whom believe it restricts their rights to free assembly.

A. Sentence 1
B. Sentence 2
C. Sentence 3
D. Sentence 4

Let's get some fitness in! Go to page 169 to try some fitness activities.

DAY 2
WEEK 2

ELA
GRAMMAR

Directions: Each sentence below contains an error related to either clarity, cohesion, or punctuation. Select the option that best corrects the error in each sentence.

Question 1:

Original: While the scientists conducted the experiment, which took over six hours they documented every significant change.

A. While the scientists conducted the experiment; which took over six hours, they documented every significant change.

B. While the scientists conducted the experiment, which took over six hours, they documented every significant change.

C. While, the scientists conducted the experiment, which took over six hours they documented every significant change.

D. While the scientists conducted the experiment which took over six hours; they documented every significant change.

Question 2:

Original: The director's new film "Lost Horizons" is set to premiere next month, critics are eager to review it.

A. The director's new film, "Lost Horizons" is set to premiere next month, critics are eager to review it.

B. The director's new film "Lost Horizons" is set to premiere next month; critics are eager to review it.

C. The director's new film "Lost Horizons," is set to premiere next month: critics are eager to review it.

D. The director's new film, "Lost Horizons," is set to premiere next month; critics are eager to review it.

Question 3:

Original: The author wrote her memoirs they span her childhood through her early career in the 1980s.

A. The author wrote her memoirs; they span her childhood through her early career in the 1980s.

B. The author wrote her memoirs they span, her childhood through her early career in the 1980s.

C. The author wrote her memoirs, they span her childhood through her early career in the 1980s.

D. The author wrote, her memoirs they span her childhood through her early career in the 1980s.

Question 4:

Original: Without hesitation, the player sprinted down the court he shot the winning basket as the buzzer sounded.

A. Without hesitation, the player sprinted down the court, he shot the winning basket as the buzzer sounded.

B. Without hesitation, the player sprinted down the court; he shot the winning basket as the buzzer sounded.

C. Without hesitation the player sprinted down the court he shot the winning basket as the buzzer sounded.

D. Without hesitation; the player sprinted down the court, he shot the winning basket as the buzzer sounded.

Question 5:

Original: The committee agreed they should meet more often to discuss progress, however, they did not set a regular schedule.

A. The committee agreed they should meet more often to discuss progress however they did not set a regular schedule.

B. The committee agreed; they should meet more often to discuss progress, however, they did not set a regular schedule.

C. The committee agreed that they should meet more often to discuss progress; however, they did not set a regular schedule.

D. The committee agreed they should meet more often to discuss progress; however they did not set a regular schedule.

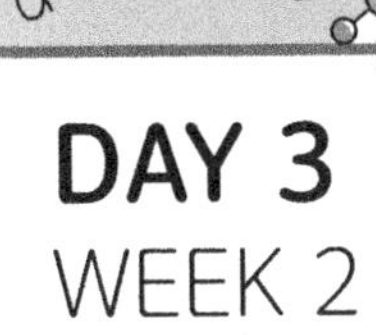

DAY 3
WEEK 2

MATH
SYSTEMS OF LINEAR EQUATIONS

OVERVIEW:

Systems of linear equations are a set of two or more equations that contain two or more variables. The solution to a system of linear equations is the set of values for each variable that makes all the equations in the system true. Let's review the methods used to solve systems of linear equations by the substitution method or the elimination method. You should have covered this topic already during 10th grade so this is a quick refresher!

The **Substitution Method** involves solving one equation for one variable in terms of the other variable, and then substituting the expression into the other equation to eliminate one variable. The steps to solve a system of linear equations by substitution are as follows:

Step 1: Solve one equation for one variable in terms of the other variable.

Step 2: Substitute the expression into the other equation to eliminate one variable.

Step 3: Solve for the remaining variable.

Step 4: Substitute the value of the remaining variable into one of the original equations to find the value of the other variable.

Step 5: Check that the values satisfy both equations.

Example:

$2x + 3y = 7$

$x - y = 1$

Step 1: Solve one of the equations for one of the variables.

We can solve the second equation for x by adding y to both sides:

$x - y = 1$

$x = y + 1$

Step 2: Substitute the expression obtained in Step 1 into the other equation.

We can substitute $y + 1$ for x in the first equation:

$2x + 3y = 7$

$2(y + 1) + 3y = 7$

Step 3: Solve for the remaining variable.

Simplifying the equation obtained in Step 2 gives:

$2y + 2 + 3y = 7$

$5y + 2 = 7$

$5y = 5$

$y = 1$

Step 4: Substitute the value obtained in Step 3 into one of the original equations to solve for the other variable.

We can substitute $y = 1$ into the second equation to solve for x:

$x - y = 1$

$x - 1 = 1$

$x = 2$

Step 5: Check the solution by substituting the values obtained in Steps 3 and 4 into one of the original equations.

We can substitute $x = 2$ and $y = 1$ into one of the two original equations to verify that the solution is correct:

$2x + 3y = 7$

$2(2) + 3(1) = 7$

$4 + 3 = 7$ CORRECT

Therefore, the solution to the system of linear equations is $x = 2$, $y = 1$. We can write the answer also in (x, y) coordinate form which is $(2, 1)$.

MATH

SYSTEMS OF LINEAR EQUATIONS

Example 2:

$3x + 4y = -5$

$2x - 3y = 8$

Let's solve for x in the second equation.

$2x - 3y = 8$

$2x = 3y + 8$

$x = \frac{3}{2}y + 4$

Now, we will **substitute** this x value for the x in the first equation.

$3x + 4y = -5$

$3\left(\frac{3}{2}y + 4\right) + 4y = -5$

Let's go ahead and solve.

$\frac{9}{2}y + 12 + 4y = -5$

$\frac{17}{2}y = -17$

$y = -2$

We know that $y = -2$ so we can plug that into either of the two original equations to figure out what x is.

$2x - 3(-2) = 8$

$2x + 6 = 8$

$2x = 2$

$x = 1$

The solution is $x = 1$ and $y = -2$. You can double-check the solution by plugging these variables back into one of the original equations. We can write the answer also in (x, y) coordinate form which is $(1, -2)$.

Your turn to practice solving systems of linear equations using the substitution method!

DAY 3
WEEK 2

MATH
SYSTEMS OF LINEAR EQUATIONS

Solve these questions using the substitution method. Do NOT use a calculator. Double check your work by plugging one of the values back into the original equations.

1. $-16x + 9y = -229$
   $-2x + y = -33$

2. $-41x + y = -123$
   $51x - 28y = 153$

3. $33x + y = 5$
   $20x + 3y = 15$

4. $x + 5y = 40$
   $-23x - 26y = 148$

5. $12x + 4y = 0$
   $-95x + y = 98$

6. $-42x + 48y = 144$
   $x + 5y = 230$

7. $17x + 85y = -289$
   $x + 3y = -3$

8. $8x + y = 123$
   $-57x + 24y = -36$

9. $97x + 23y = 36$
   $66x + y = -122$

10. $17x + y = 227$
    $-14x + 22y = -50$

11. $x + 75y = 103$
    $12x - 70y = 266$

12. $-96x + 87y = 225$
    $-12x + y = 275$

13. $92x - 88y = 72$
    $x + 52y = -264$

14. $7x - 23y = 28$
    $5x + y = 20$

15. $-71x + 7y = 206$
    $-x + y = -62$

Let's get some fitness in! Go to page 169 to try some fitness activities.

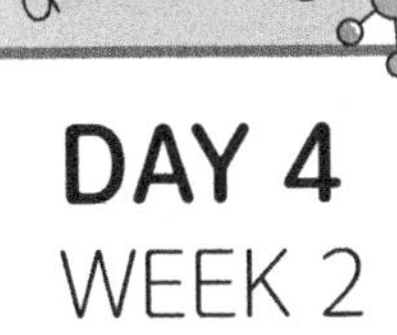

DAY 4
WEEK 2

MATH
SYSTEMS OF LINEAR EQUATIONS

OVERVIEW:

We can solve systems of linear equations using another method called **Elimination Method**.

This method involves adding or subtracting the equations to eliminate one variable and solve for the other.

The general idea of the elimination method is to manipulate the equations so that one variable has opposite coefficients in each equation, and then add or subtract the equations to eliminate that variable. This results in an equation in one variable, which can be solved to obtain the value of that variable. The value of the other variable can then be obtained by substituting the value obtained into one of the original equations.

Let's consider an example of a system of two linear equations:

$3x + 2y = 9$ (Equation 1)

$2x + 6y = 6$ (Equation 2)

Step 1: Choose a variable to eliminate. For this example, let's go ahead and eliminate the variable "x". How can we do this? We can manipulate Equation 1 by multiplying the entire equation by -2 **and** we can multiply the Equation 2 by 3.

$$-2 \quad (3x + 2y = 9)$$
$$3 \quad (2x + 6y = 6)$$

That gives us

$$-6x - 4y = -18$$
$$+6x - 18y = 18$$

This is great because we just eliminated the variable x. We are just left with $-22y = 0$. Let's solve for y, by dividing both side by -22.

y = 0.

Since we now know that $y = 0$, we can plug that value into any of our original equations. Let's plug that into our original Equation 1.

$3x + 2(0) = 9$

$3x + 0 = 9$

$3x = 9$

x = 3

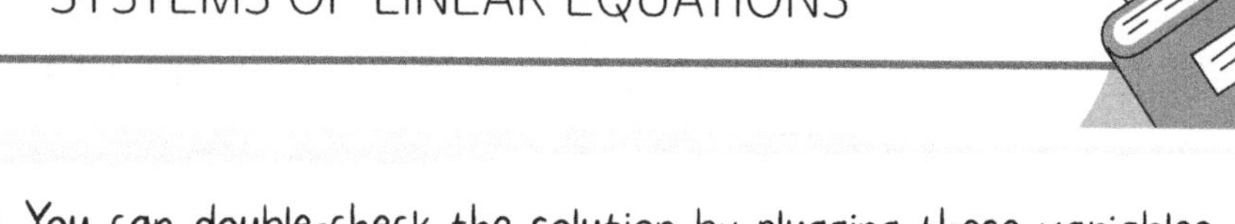

The solution is $x = 3$ and $y = 0$. You can double-check the solution by plugging these variables back into one of the original equations. We can write the answer also in (x, y) coordinate form which is $(3, 0)$.

Let's take a look at one more practice example using the Elimination Method.

$-x + 5y = 8$

$3x + 7y = -2$

We can easily get rid of the variable x if we manipulate Equation I by multiplying by 3 and then adding the two equations together.

$$\begin{cases} -x + 5y = 8 \\ 3x + 7y = -2 \end{cases} \overset{\times 3}{\Rightarrow} \begin{cases} -3x + 15y = 24 \\ 3x + 7y = -2 \end{cases}$$

$$\begin{array}{r} -3x + 15y = 24 \\ +\quad 3x + 7y = -2 \\ \hline 22y = 22 \end{array}$$

$$\frac{22y}{22} = \frac{22}{22}$$

$$y = 1$$

We know $y = 1$ so let's plug that back into one of the original equations and solve for x.

$-x + 5(1) = 8$

$-x + 5 = 8$

$-x = 3$

$x = -3$

The solution is $x = -3$ and $y = 1$. You can double-check the solution by plugging these variables back into one of the original equations. We can write the answer also in (x, y) coordinate form which is $(-3, 1)$.

Your turn to practice solving systems of linear equations using the elimination method!

DAY 4
WEEK 2

MATH
SYSTEMS OF LINEAR EQUATIONS

Solve these questions using the elimination method. Do NOT use a calculator. Double check your work by plugging one of the values back into the original equations.

1. $-5x - 4y = -2$
$-x + 2y = 8$

2. $-x + 4y = 7$
$5x - 8y = -11$

3. $x + 3y = -7$
$-4x - y = 17$

4. $5x + 2y = -10$
$x - y = -2$

5. $x - 8y = 10$
$-6x - 4y = -8$

6. $-2x + 4y = 0$
$-x - 8y = -10$

7. $6x + 6y = 6$
$-12x - 4y = 4$

8. $-2x - 4y = 10$
$4x - 5y = -7$

9. $-8x + 5y = -5$
$4x + 3y = -3$

10. $-3x + 12y = -6$
$x + 6y = 2$

11. $x - 2y = 10$
$-2x - 4y = -4$

12. $2x - 12y = 0$
$-4x + 6y = -18$

13. $-6x - y = -18$
$x + 2y = 3$

14. $-6x + 2y = -8$
$-2x - y = -11$

15. $-x + 4y = -11$
$6x - 8y = 18$

Let's get some fitness in! Go to page 169 to try some fitness activities.

DAY 5
WEEK 2

MATH
COMPOUND INTEREST

OVERVIEW:

Compound interest is a way of calculating interest where you earn interest not only on the initial amount of money (the principal) but also on the interest that has been added to your initial investment over time.

The formula to calculate compound interest is:

$$A = P\left(1 + \frac{r}{n}\right)^{nt}$$

where:

A is the amount of money accumulated after n years, including interest.

P is the principal amount (the original sum of money).

r is the annual interest rate (decimal).

n is the number of times that interest is compounded per year.

t is the time the money is invested or borrowed for, in years.

Lete's take a look at an example. Suppose you save $1,000 at an interest rate of 5% per year, compounded annually, for 3 years. How much would you have after 3 years?

Write down the known variables.

P = $1,000

r = 0.05 (r represents the interest rate which is expressed as a percent. 5% is 0.05 in decimal form)

n = 1 (compounded yearly)

t = 3 years

$$A = 1000\left(1 + \frac{0.05}{1}\right)^{1 \times 3} = 1000 \times 1.157625 = \$1,157.63$$

You would have $1,157.63 after 3 years.

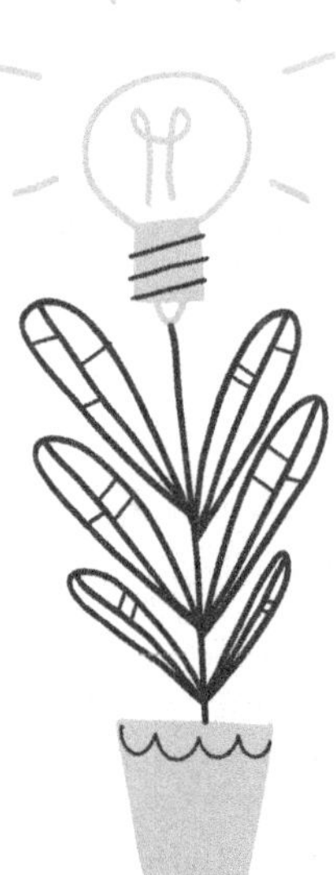

DAY 5
WEEK 2

MATH
COMPOUND INTEREST

Let's try this example together.

Eduard invests **\$8,271** in a retirement account with a fixed annual interest rate of **5%** compounded continuously. What will the account balance be after **14** years?

Write down the known variables.

$P = \$8,271$

$r = 0.05$

$n = 1$ (compounded yearly)

$t = 14$

$A = ?$

Plug all the numbers back into our equation to get

$$A = 8271\left(1 + \frac{0.05}{1}\right)^{1 \times 14}$$

$$A = 8271 \times 1.9799 = \$16,375.75$$

Eduard will have an account balance of **\$16,375.75** after **14** years.

Your turn to practice solving compound interest word problems!

Notes:

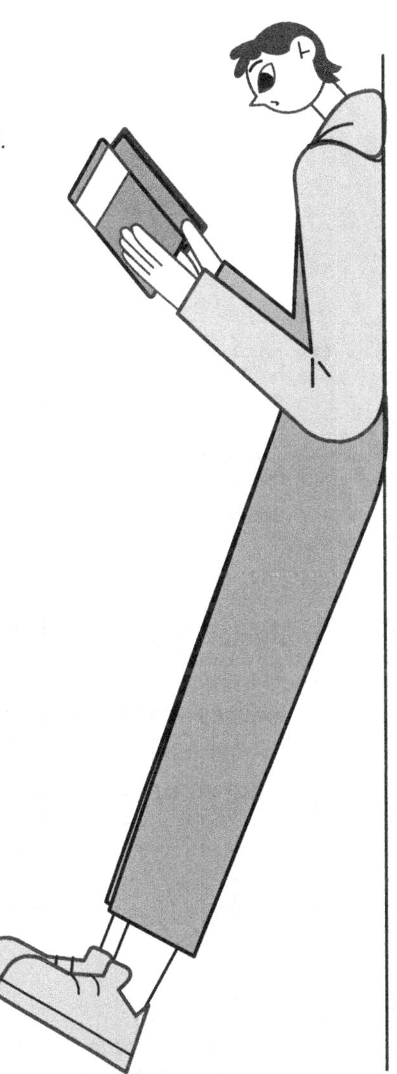

MATH

COMPOUND INTEREST

Solve compound interest word problems.

1. Stacy invests $8,845 in a savings account with a fixed annual interest rate of 2% compounded continuously. What will the account balance be after 4 years?

2. Lena invests $3,539 in a savings account with a fixed annual interest rate of 9% compounded continuously. What will the account balance be after 7 years?

3. Ryan invests $7,176 in a savings account with a fixed annual interest rate of 3% compounded continuously. What will the account balance be after 8 years?

4. Michelle invests $8,079 in a savings account with a fixed annual interest rate of 7% compounded continuously. What will the account balance be after 5 years?

5. James invests $7,540 in a savings account with a fixed annual interest rate of 9% compounded continuously. What will the account balance be after 10 years?

6. Arjun invests $2,246 in a savings account with a fixed annual interest rate of 3% compounded 12 times per year. What will the account balance be after 9 years?

7. Jimmy invests $3,893 in a savings account with a fixed annual interest rate of 3% compounded 12 times per year. What will the account balance be after 6 years?

8. Rob invests $3,575 in a retirement account with a fixed annual interest rate of 7% compounded 4 times per year. What will the account balance be after 16 years?

9. Vlad invests $4,815 in a savings account with a fixed annual interest rate of 2% compounded 6 times per year. What will the account balance be after 4 years?

10. Jill invests $6,877 in a retirement account with a fixed annual interest rate of 4% compounded 4 times per year. What will the account balance be after 13 years?

Let's get some fitness in! Go to page 169 to try some fitness activities.

DAY 6
WEEK 2

ELA
MODIFIERS, DANGLING MODIFIERS, AND CLAUSE STRUCTURES

OVERVIEW:

Modifiers:

Modifiers are words, phrases, or clauses that provide additional information about other elements in a sentence. They describe, qualify, or limit these elements, enhancing detail and specificity in language. Modifiers can be adjectives or adverbs, as well as modifying phrases or clauses. The key to using modifiers effectively is to place them near the word or phrase they are intended to modify, to avoid confusion.

- Adjective Modifiers: Adjectives modify nouns or pronouns. For example, in the sentence "The green apple is ripe," "green" modifies "apple."
- Adverbial Modifiers: Adverbs can modify verbs, adjectives, or other adverbs. For instance, "He drives very fast," where "very" modifies "fast."

A **dangling modifier** occurs when the modifier does not logically attach to the word or words it's supposed to modify. This usually happens because the word that the modifier is meant to describe is missing from the sentence, leaving the modifier to appear to modify the wrong word or phrase. Dangling modifiers can lead to unclear or absurd sentences.

Examples of a dangling modifier:

- "Having finished the assignment, the TV was turned on."

It sounds as though the TV finished the assignment. This sentence would be clear and logical if we correct it to "Having finished the assignment, she turned on the TV."

- "Without knowing his name, it was difficult to introduce him."

In this sentence it's unclear who does not know the name. A better sentence ewould be "Because she did not know his name, it was difficult to introducc him."

Clause Structures:

Clauses are groups of words that contain a subject and a verb. Clauses can be independent, meaning they can stand alone as a sentence, or dependent, meaning they cannot stand alone and are used as part of a larger sentence.

- Independent Clauses: "She enjoys painting," is a complete thought and can stand alone as a sentence.

- Dependent Clauses: "Although she enjoys painting," is an incomplete thought and needs an independent clause to form a complete sentence, like "Although she enjoys painting, she does not want to do it professionally."

Effective writing involves using clause structures to create complex, nuanced sentences. **Here are some tips.**

- Ensure that modifiers are placed close to the words they modify.
- Check that the intended subject of a modifier is present in the sentence to avoid dangling modifiers.
- Use correct punctuation and conjunctions to link clauses appropriately, maintaining clarity and enhancing the flow of information.

Great! Try these practice questions.

Directions: Each sentence below contains an error related to misplaced or dangling modifiers, or incorrect clause usage. Choose the option that correctly repositions or revises the modifier or clause for clarity.

Question 1:

Original: Running through the park, a squirrel startled her.

A. Running through the park, she was startled by a squirrel.
B. She startled a squirrel running through the park.
C. While she ran through the park, a squirrel was startling her.
D. She was running through the park when a squirrel startled her.

Question 2:

Original: To win the championship, the last game was crucial.

A. To win the championship, they needed to win the last game.
B. The last game was crucial to winning the championship.
C. The championship required winning the last game to win.
D. Winning the last game, the championship was theirs.

Question 3:

Original: Having finished the project, the celebration was enjoyed by the team.

A. The celebration was enjoyed by the team, having finished the project.
B. Having finished the project, the team enjoyed the celebration.
C. The team enjoyed the celebration after having finished the project.
D. The team, having finished the project, enjoyed the celebration.

Question 4:

Original: The hiker, exhausted from the long trek, the campsite was a welcome sight.

A. Exhausted from the long trek, the campsite was a welcome sight for the hiker.
B. The campsite was a welcome sight for the hiker, exhausted from the long trek.
C. Exhausted from the long trek, the hiker found the campsite a welcome sight.
D. The hiker found the campsite, exhausted from the long trek, a welcome sight.

Question 5:

Original: Waiting for the bus, the rain started to pour.

A. As I was waiting for the bus, the rain started to pour.
B. The rain started to pour, waiting for the bus.
C. Waiting for the bus, I was caught in the pouring rain.
D. The rain, waiting for the bus, started to pour.

Let's get some fitness in! Go to page 169 to try some fitness activities.

DAY 7 WEEK 2

Logic Games

INTERESTING

1. All Wugs are Wogs. Some Wogs are Wigs. Which of the following must be true?

A. All Wugs are Wigs.

B. Some Wugs are Wigs.

C. No Wugs are Wigs.

D. Some Wigs are Wugs.

· o o o ·

2. If APPLE is coded as BQQMF, how would you code BANANA?

· o o o ·

3. A fair die is rolled twice. What is the probability of getting a sum of 7 or 11?

A. $\frac{1}{9}$ **B.** $\frac{1}{6}$ **C.** $\frac{2}{9}$ **D.** $\frac{1}{3}$

· o o o ·

4. The sum of three consecutive even numbers is 48. What is the largest of these numbers?

A. 12

B. 14

C. 16

D. 18

· o o o ·

5. The ratio of the number of boys to the number of girls in a class is 3:2. If there are 30 boys in the class, how many students are there in total?

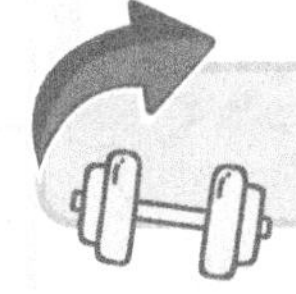

Let's get some fitness in! Go to page 169 to try some fitness activities.

WEEK 3

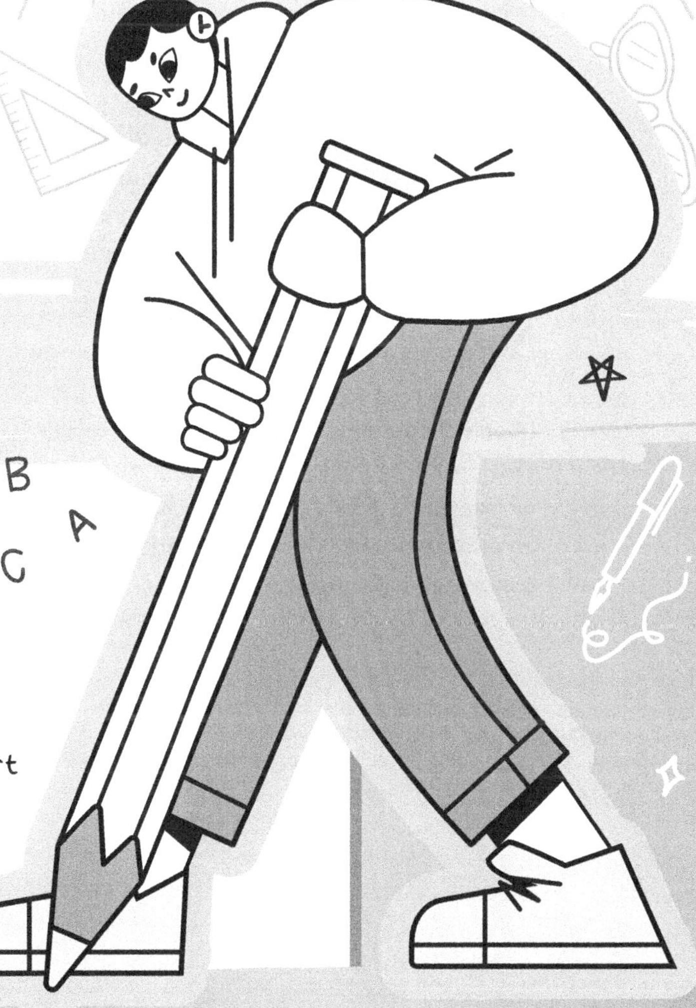

GRADE 10-11

This week, you will explore the concepts of themes and motifs in literature, focusing on how these elements contribute to the overall meaning and message of a story. You will also learn about the Unit Circle and how to convert degrees to radians and radians to degrees. Additionally, this week you will start preparing for the SAT ELA section with practice questions.

DAY 1
WEEK 3

ELA
THEMES & MOTIFS

OVERVIEW:

In literary studies, the exploration of themes and motifs provides a framework for deeper understanding and critical analysis of texts across genres and historical periods. Themes are the **central**, **underlying**, and often universal ideas explored in a literary work. They are the fundamental and often universal ideas explored through the narrative. Motifs, on the other hand, are recurring structures, contrasts, or literary devices that can help to develop and inform the text's major themes.

Understanding Themes

A theme is not just a topic, but an insight or opinion expressed about that topic. For instance, in F. Scott Fitzgerald's "The Great Gatsby," one of the prominent themes is the critique of the American Dream, exploring how the pursuit of material wealth and social status leads to moral decay and personal dissatisfaction.

Themes in literature are not always explicitly stated; rather, they are often implied and must be extracted by analyzing characters' actions, dialogues, and the narrative's progression. Identifying themes involves recognizing patterns in how characters react to their circumstances and each other, and how these reactions reflect broader societal issues.

Motifs

Motifs support and enrich the understanding of themes by highlighting recurring elements that are significant to the text's message. These can include imagery, symbols, sounds, or actions. For example, the motif of "sight and blindness" in Shakespeare's "King Lear" emphasizes the theme of understanding and recognition, portraying literal and metaphorical blindness to truth.

DAY 1
WEEK 3

ELA
THEMES & MOTIFS

Directions: Read the following passage carefully and then answer the multiple-choice questions that follow.

In the quiescent hours of the night, when the world lay enveloped in a shroud of darkness, the old man sat by the fireplace, his weathered hands clasped around a tattered journal. The flickering flames cast an ethereal glow upon his face, illuminating the deep crevices that mapped his life's journey. As he turned the yellowed pages, a wistful smile played upon his lips, memories of a distant past flooding his mind.

He had lived a life of profound experiences, each one etched indelibly into the fabric of his being. From the tender innocence of childhood to the tempestuous passions of youth, he had embraced every moment with an unwavering fervor. Love had painted his world in vivid hues, while loss had taught him the fragility of existence. Through triumphs and tribulations, he had danced to the cadence of life's symphony, always striving to find meaning amidst the chaos.

As he delved deeper into the journal, the old man's eyes misted over, for within those pages lay a testament to the ephemeral nature of time. The faces that smiled back at him from faded photographs were but ghostly remnants of a bygone era, reminders of the relentless march of years. Yet, even as the sands of time slipped through his fingers, he found solace in the realization that his legacy would endure, woven into the tapestry of the lives he had touched.

With a heavy sigh, the old man closed the journal, his heart brimming with a bittersweet mixture of nostalgia and gratitude. He knew that the final chapter of his story was drawing to a close, but he faced the inevitable with a sense of peace, knowing that he had lived a life worth remembering. As the embers of the fire dwindled to ash, he closed his eyes, content in the knowledge that his spirit would live on, forever etched in the annals of time.

Questions:

1. Which of the following best describes the main theme of the passage?

A. The importance of keeping a journal
B. The fleeting nature of youth and love
C. The inevitability of change and the passage of time
D. The power of memories to provide comfort in old age

2. The old man's reflections on his life experiences suggest that:

A. Life is a series of disconnected events
B. One should avoid forming deep attachments
C. Life is a meaningful journey filled with both joys and sorrows
D. The past is best forgotten and left behind

3. The metaphor of the "sands of time" in the passage symbolizes:

A. The old man's fading memories
B. The transient nature of human existence
C. The physical deterioration that comes with age
D. The cyclical patterns of life and death

4. The old man's sense of peace at the end of the passage arises from:

A. His acceptance of the inevitability of death
B. His satisfaction in knowing that the memories of his life are preserved in his journal
C. His relief at finally finishing his journal
D. His hope that he will be reunited with loved ones in the afterlife

Notes:

Let's get some fitness in! Go to page 169 to try some fitness activities.

Directions: Read the following passage carefully and then answer the multiple-choice questions that follow.

The city streets pulsated with an electric energy, a cacophony of sounds and sights that assaulted the senses. Amidst the chaos, a young woman navigated the labyrinthine paths, her eyes fixed on the horizon, her heart filled with an insatiable hunger for something more. She had grown weary of the mundane existence that had been thrust upon her, the endless cycle of monotony that threatened to erode her very essence.

In the quiet moments, when the world seemed to pause for a fleeting breath, she would close her eyes and imagine a life beyond the confines of the concrete jungle. She dreamed of rolling hills, of verdant landscapes untouched by the relentless march of progress. It was in these stolen instances that she felt truly alive, her spirit soaring on the wings of possibility.

Yet, reality always came crashing back, a harsh reminder of the inescapable truth. She was bound by the chains of societal expectations, her dreams relegated to the realm of fantasy. The weight of responsibility pressed down upon her, a suffocating force that threatened to extinguish the flames of her passion.

But deep within her heart, a flicker of hope refused to be extinguished. She knew that somewhere, beyond the boundaries of her current existence, lay a world waiting to be explored. And so, she persevered, her steps growing ever more determined, her resolve unwavering in the face of adversity.

As she walked, the city began to transform before her eyes. The towering skyscrapers became mountains to be scaled, the winding streets rivers to be crossed. Each obstacle became a challenge to be overcome, a testament to her indomitable spirit. And with every step, she moved closer to her dream, to the life she knew she was meant to live.

In the end, she emerged victorious, her soul forged in the crucible of her struggles. She had broken free from the shackles of conformity, her spirit soaring on the winds of change. And as she stood atop the precipice of her new existence, she knew that she had finally found her true purpose, her reason for being. With a smile on her face and a fire in her heart, she stepped forward, ready to embrace the endless possibilities that lay ahead.

Questions:

1. The main theme of the passage can be best described as:

   A. The importance of conforming to societal expectations
   B. The power of hope and perseverance in the face of adversity
   C. The futility of dreaming and the acceptance of one's fate
   D. The corrupting influence of the city on the human spirit

2. The young woman's dreams of a life beyond the city represent:

   A. A desire for escapism and avoidance of responsibility
   B. A longing for a simpler, more fulfilling existence
   C. A rejection of modern technology and progress
   D. A fear of the unknown and the unfamiliar

3. The transformation of the city in the young woman's eyes symbolizes:

   A. The literal change in her physical surroundings
   B. The shifting nature of reality and perception
   C. Her inner growth and changing perspective on life's challenges
   D. The illusory nature of her dreams and aspirations

4. The young woman's ultimate triumph at the end of the passage suggests:

   A. The importance of compromising one's dreams for practicality
   B. The need to rebel against society's norms and expectations
   C. The power of the individual to shape their own destiny
   D. The inevitability of success for those who dare to dream

Let's get some fitness in! Go to page 169 to try some fitness activities.

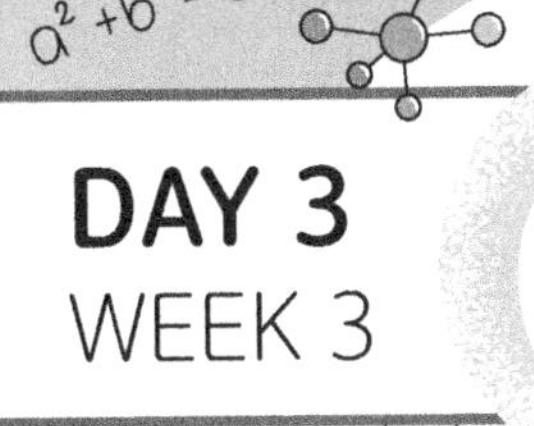

DAY 3
WEEK 3

MATH
UNIT CIRCLE

OVERVIEW:

The concept of the **unit circle** is fundamental to trigonometry, bridging the gap between the linear world of algebra and the cyclical patterns observed in trigonometry. At its core, the unit circle is a simple geometric shape - a circle with a radius of one - but its implications and applications in mathematics, and specifically in trigonometry, are profound.

Imagine a circle, centered at the origin of a coordinate plane, with a radius of exactly one unit. This circle is aptly named the **"unit circle."** While the idea might seem basic, as we journey around this circle, we will discover that **every point** on its circumference has a unique relationship to the angles it creates and the lengths of the sides of the triangles we can form inside the circle.

These relationships give rise to the fundamental trigonometric functions: sine, cosine, and tangent.

One of the primary reasons the unit circle is so invaluable is its ability to allow us to **define** trigonometric functions for **all real numbers**, not just specific angles.

Take a look at the diagram below. This is a Unit Circle. As you can see, the **radius** is 1 and the center is at the origin (0,0).

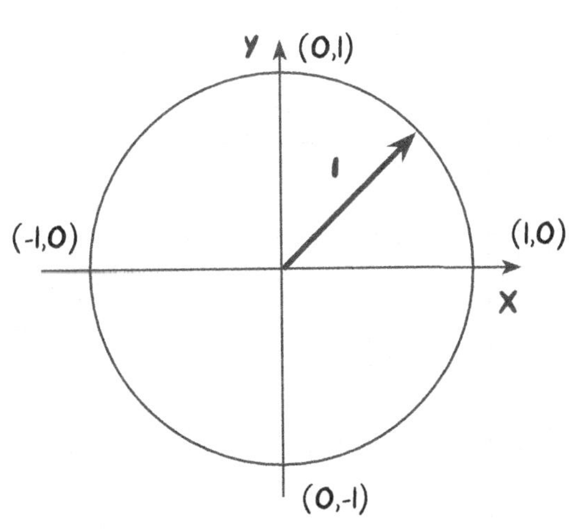

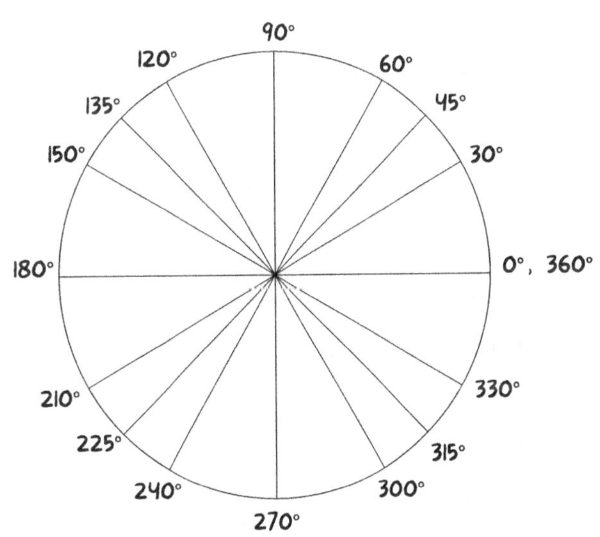

You know that a circle has 360°.

Arcs on the Unit Circle

When you hear the phrase "a piece of the pie," you can think of an arc as representing the crusty edge of that pie slice. In simple words, an arc is a segment or a portion of a circle's circumference.

Now, consider the unit circle, which is centered at the origin $(0, 0)$ of a coordinate plane and has a radius of exactly one unit. Every point on this circle is exactly one unit away from the center.

When we talk about an angle formed in the unit circle, we generally refer to the angle formed between the x-axis and a line segment (or radius) drawn from the center of the circle to a point on the circle. As this line segment (or radius) sweeps or rotates from its initial position, it "cuts out" or delineates an arc on the circle. The size of this arc is directly related to the size of the angle.

Measuring Arcs

In trigonometry, one of the first skills you need to master as a student is to convert degrees to radians, and radians to degrees which you will practice this week.

Arcs on the unit circle can be measured in two common ways:

1. Degrees: This is the measurement you're most familiar with. A full circle contains $360°$, so if a radius of the unit circle rotates $90°$ from the positive x-axis, the arc it sweeps out is $\frac{1}{4}$ of the circle's circumference.
2. Radians: This is a more "natural" way to measure angles when working in the context of circles, especially the unit circle. A **full circle** is 2π radians.

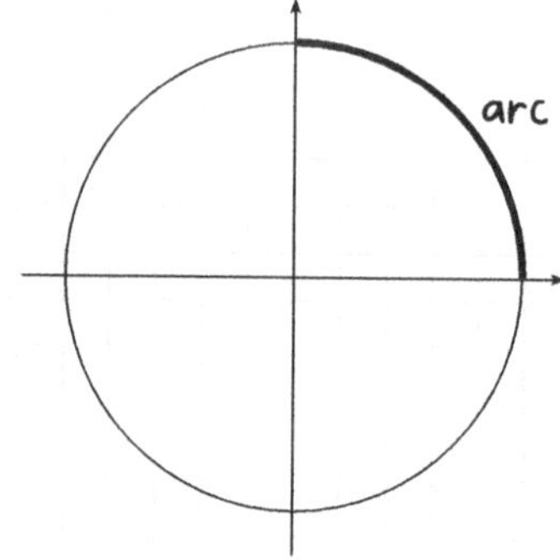

Now that you know in a **full circle** there are 2π radians, how many radians are in half a circle?

In half a circle, there are π radians, which is also $180°$.

DAY 3
WEEK 3

MATH
UNIT CIRCLE

Now that we know a few fundamental information of a Unit Circle, let's take a close look at the diagram below. This should make sense to you.

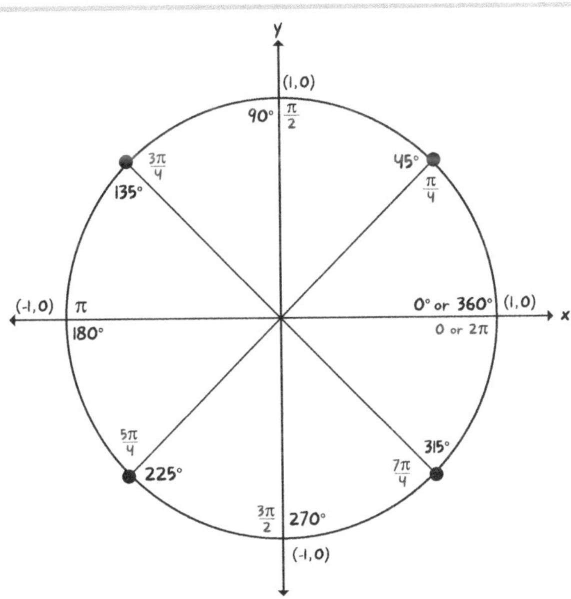

This is a Unit Circle diagram that is divided into 45° increments. We see at 360° that we have 2π radians. At the 180° position, we see π radians. This is information we already know.

How about 90°? What is the radian equivalent for 90°? Since π radians = 180°, we can just divide both sides by 2 to get $\frac{\pi}{2}$ = 90°.

How about 45°? What is the radian equivalent for 45°? Since π radians = 180°, we can just divide both sides by 4 to get $\frac{\pi}{4}$ = 45°.

Here is a chart below of common degrees and radian conversion.

Degrees	Radians
0	0
30	$\frac{\pi}{6}$
45	$\frac{\pi}{4}$
60	$\frac{\pi}{3}$
90	$\frac{\pi}{2}$
180	π
270	$\frac{3\pi}{2}$
360	2π

While you **do not need** to memorize this entire chart, it is helpful to remember a few of these to make your calculations faster. We can always find out the answer since we know that π radians = 180°.

DAY 3 WEEK 3

MATH
UNIT CIRCLE

Answer the following questions.

1. How many degrees are equivalent to π radians?

A. 90° **C.** 270°
B. 180° **D.** 360°

2. Which angle measure is equivalent to one full rotation around the unit circle?

A. 180° **C.** π radians
B. 270° **D.** 2π radians

3. What is the relationship between degrees and radians?

A. π radians = 180°
B. 1 radian = 180°
C. π radians = 90°
D. 1 radian = 360°

4. If an angle in the unit circle is $\frac{\pi}{4}$ radians, how many degrees is this?

A. 45° **C.** 135°
B. 90° **D.** 180°

5. How many radians are in three-quarters of a full rotation around the circle?

A. $\frac{3\pi}{4}$ radians **C.** $\frac{3\pi}{2}$ radians
B. $\frac{5\pi}{4}$ radians **D.** $\frac{7\pi}{4}$ radians

6. Which of the following is an acute angle?

A. π radians
B. 2π radians
C. $\frac{3\pi}{2}$ radians
D. $\frac{\pi}{6}$ radians

7. Each radian is _____ degrees.

A. $\frac{90}{\pi}$ **C.** $\frac{180}{\pi}$
B. 90 **D.** 45

8. Which of the following is an obtuse angle?

A. 30°
B. $\frac{\pi}{4}$ radians
C. $\frac{\pi}{3}$ radians
D. 94°

Let's get some fitness in! Go to page 169 to try some fitness activities.

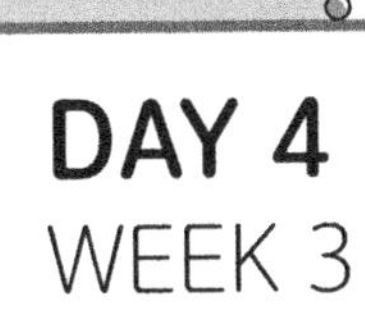

DAY 4
WEEK 3

MATH
CONVERTING DEGREES TO RADIANS

OVERVIEW:

We learned that

π radians = 180° and 2π radians = 360°

Using this information, 1 Radian is equal to how many degrees?

We know π radians = 180°

To solve for 1 radian, we need to divide π on both sides.

$$\frac{\cancel{\pi}\text{ radians}}{\cancel{\pi}} = \frac{180^\circ}{\pi}$$

We see that 1 radian = $\frac{180^\circ}{\pi}$.

If you use a calculator and divide 180 degrees by π, **you will get approximately 57.2958...°**

You do **not** need to remember this information, but you **must know** that 1 radian = $\frac{180^\circ}{\pi}$

Knowing this helps us to go from **radians to degrees** and **degrees to radians.**

Here is the rule to convert **degrees to radians.**

To convert degrees to radians, multiply by $\frac{\pi}{180}$

Example:

Convert 65° to radians.

We know the rule "To convert degrees to radians, multiply by $\frac{\pi}{180}$".

$65 \times \frac{\pi}{180} = \frac{65\pi}{180}$

We can simplify this further by dividing the top and bottom by 5.

$\frac{65\pi \div 5}{180 \div 5} = \frac{13\pi}{36}$

The answer is $\frac{13\pi}{36}$ radians.

Your turn to practice converting degrees to radians!

DAY 4
WEEK 3

MATH
CONVERTING DEGREES TO RADIANS

Convert the given angle from degrees to radians.

1. 330°
2. 125°
3. 190°
4. 105°
5. 65°
6. 255°
7. 70°
8. 270°
9. 55°
10. 220°
11. 5°
12. 250°
13. 40°
14. 205°
15. 90°
16. 340°
17. 290°
18. 45°
19. 200°
20. 310°
21. 35°
22. 235°
23. 300°
24. 245°
25. 110°
26. 175°
27. 80°
28. 355°
29. 160°
30. 130°

Notes:

Let's get some fitness in! Go to page 169 to try some fitness activities.

DAY 5
WEEK 3

MATH
CONVERTING RADIANS TO DEGREES

OVERVIEW:

Here is the rule to convert **radians to degrees.**

To convert radians to degrees, multiply by $\frac{180°}{\pi}$

Example 1:

Convert $\frac{11\pi}{18}$ radians to degrees.

We know the rule "To convert radians to degrees, multiply by $\frac{180°}{\pi}$".

$$\frac{11\pi}{18} \times \frac{180°}{\pi} = \frac{1{,}980°\pi}{18\pi}$$

We can simplify this further. Notice that the π's cancel out because there is a π on the numerator and also on the denominator. We are only left with $\frac{1{,}980}{18}$ which gives us 110°.

The answer is 110°.

Example 2:

Convert $\frac{17\pi}{36}$ radians to degrees.

We know the rule "To convert radians to degrees, multiply by $\frac{180°}{\pi}$".

$$\frac{17\pi}{36} \times \frac{180°}{\pi} = \frac{3{,}060°\pi}{36\pi}$$

We can simplify this further. Notice that the π's cancel out because there is a π on the numerator and also on the denominator. We are only left with $\frac{3{,}060}{36}$ which gives us 85°.

The answer is 85°.

Your turn to practice converting radians to degrees!

DAY 5
WEEK 3

MATHEMATICS
Rhetorical Strategies in Historical Fiction

Convert the given angle from radians to degrees.

1. $\frac{47\pi}{36}$
2. $\frac{3\pi}{2}$
3. $\frac{35\pi}{36}$
4. $\frac{\pi}{36}$
5. $\frac{59\pi}{36}$
6. $\frac{2\pi}{3}$
7. $\frac{19\pi}{12}$
8. $\frac{19\pi}{18}$
9. $\frac{25\pi}{18}$
10. $\frac{13\pi}{18}$
11. $\frac{13\pi}{36}$
12. $\frac{\pi}{4}$
13. $\frac{\pi}{6}$
14. $\frac{5\pi}{3}$
15. $\frac{5\pi}{36}$
16. $\frac{3\pi}{4}$
17. $\frac{43\pi}{36}$
18. $\frac{23\pi}{36}$
19. $\frac{53\pi}{36}$
20. $\frac{2\pi}{9}$
21. $\frac{35\pi}{18}$
22. $\frac{71\pi}{36}$
23. $\frac{\pi}{3}$
24. $\frac{\pi}{2}$
25. $\frac{29\pi}{18}$
26. $\frac{41\pi}{36}$
27. $\frac{17\pi}{18}$
28. $\frac{31\pi}{18}$
29. $\frac{\pi}{12}$
30. 2π

Let's get some fitness in! Go to page 169 to try some fitness activities.

DAY 6
WEEK 3

ELA
VOCABULARY

Directions: Read each sentence carefully and use context clues to determine the meaning of the underlined word. Choose the best definition for the underlined word from the options provided. Then, write a sentence using the word in a new context, demonstrating your understanding of its meaning and usage. You can use the Internet to look up the exact definition after answering the questions.

1. The scientist's hypothesis was met with **opprobrium** from her colleagues, who found it lacking in evidence and methodological rigor.

A. praise
B. indifference
C. curiosity
D. harsh criticism

Sentence:

....

....

....

2. The diplomat's **aplomb** in the face of the international crisis was a testament to her years of experience and unflappable demeanor.

A. uncertainty
B. self-assurance
C. indecision
D. agitation

Sentence:

....

....

....

ELA
VOCABULARY

3. The novel's **denouement** was a tour de force of literary craftsmanship, tying together seemingly disparate plot threads with breathtaking skill.

A. opening scene
B. climax
C. final resolution
D. main conflict

Sentence:

4. The politician's **grandiloquent** speeches were often short on substance, leaving voters yearning for more concrete solutions.

A. plain-spoken
B. pompous
C. articulate
D. persuasive

Sentence:

5. The politician's **pusillanimous** response to the crisis left many questioning his leadership abilities.

A. cowardly
B. bold
C. decisive
D. thoughtful

Sentence:

DAY 7
WEEK 3

SAT
ELA PREP

OVERVIEW:

Answering questions on the SAT English Language Arts section that ask you to choose the most logical and precise word or phrase to complete a text can be challenging. Here's an overview and some strategic tips to help you effectively tackle these types of questions.

These questions are designed to assess your ability to understand and manipulate language in context. The goal is to select a word or phrase that best fits semantically and syntactically within a given passage. Precision in vocabulary and context awareness are crucial for identifying the most appropriate answer.

Tips for Answering Vocabulary in Context Questions

1. Read the Sentence Carefully
   - Before looking at the answer choices, read the sentence or passage carefully to understand the overall meaning and tone. Pay special attention to the words immediately surrounding the blank, as they often provide significant clues about the correct answer.
2. Predict the Answer
   - Try to predict what type of word (noun, verb, adjective, etc.) or the specific word would logically complete the sentence before looking at the options. This helps reduce bias introduced by the answer choices.
3. Analyze Each Option
   - Evaluate each answer choice individually within the context of the sentence. Insert the word or phrase into the blank and decide if it makes sense both logically and grammatically. Consider connotations and subtle differences in meaning.
4. Look for Contradictions
   - Discard any options that contradict other information in the sentence or passage. Sometimes, wrong answers are plausible in a general sense but do not fit the specific context provided.
5. Consider Connotations and Subtleties
   - Some words may seem correct but might carry connotations that make them inappropriate in the context. Be sensitive to these nuances, as the SAT often includes such traps.

6. Use Process of Elimination

   - If you're unsure, use the process of elimination to narrow down the choices. Remove any answer that is clearly inconsistent with the tone, style, or factual content of the passage.

7. Re-read with Your Selected Answer

   - Once you select an answer, re-read the sentence with that answer in place to ensure it flows naturally and maintains the passage's intended meaning and tone.

8. Practice with Diverse Materials

   - Regularly reading and practicing with a variety of texts—literary, scientific, historical, and social sciences—can improve your ability to quickly understand context and choose appropriate words.

9. Study Vocabulary

   - Improving your vocabulary can directly help with these questions, as familiarity with the words presented as choices allows for quicker and more accurate answers.

10. Stay Calm and Manage Time

   - These questions can be tricky, and it might be tempting to spend too much time on them. Practice pacing to ensure that you have enough time for all sections of the test.

Try the following practice questions!

1. Dr. Fiona Mwangi, a leading climate scientist, emphasizes that while specific predictions are challenging, she ________ that climate change will significantly alter coastal ecosystems. This belief drives her commitment to studying sea level rise and its impacts on marine habitats.

   Which choice completes the text with the most logical and precise word or phrase?

   A. asserts
   B. imagines
   C. dismisses
   D. contradicts

Let's get some fitness in! Go to page 169 to try some fitness activities.

2. Dr. Miriam Joyce, a historian specializing in the Cold War, ________ that the proliferation of nuclear arms had a paradoxical role in maintaining peace through the threat of mutual destruction. Her publications extensively analyze this theory of deterrence.

   Which choice completes the text with the most logical and precise word or phrase?

   A. disputes
   B. maintains
   C. regrets
   D. reminisces

3. In response to rising concerns over privacy and data security, several tech companies have adopted a framework that emphasizes transparency and user control. The partnership between Civic Tech Solutions and the Digital Rights Foundation ________ this approach by developing open-source tools that allow users to monitor and manage their data more effectively.

   Which choice completes the text with the most logical and precise word or phrase?

   A. obscures
   B. reinforces
   C. contradicts
   D. supersedes

4. The rapid evolution of antibiotic-resistant bacteria poses a significant challenge to modern medicine. Dr. Hector Alvarez's research into phage therapy ________ a promising strategy to combat these superbugs by exploiting natural viral predators of bacteria.

   Which choice completes the text with the most logical and precise word or phrase?

   A. underestimates
   B. relinquishes
   C. showcases
   D. circumvents

5. Contemporary artist Maya Lin ________ the conventional boundaries between sculpture and architecture: this blending is evident in her design of the Vietnam Veterans Memorial, which integrates the landscape and viewer interaction in a reflective, abstract form.

   Which choice completes the text with the most logical and precise word or phrase?

   A. transgresses
   B. acknowledges
   C. revises
   D. transcends

WEEK 4

GRADE 10-11

This week, you will learn strategies for reading and analyzing poetry, including active reading, identifying themes, and exploring figurative language. For math, you will learn about the six trigonometric functions and conclude the week with more SAT ELA prep and vocabulary.

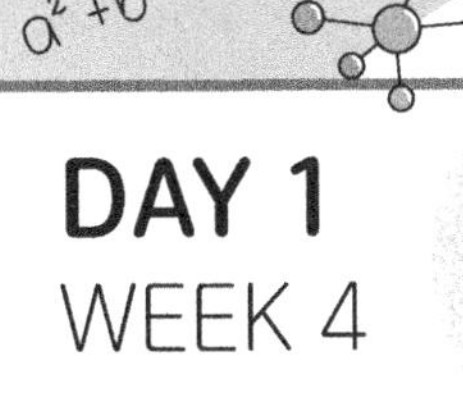

DAY 1
WEEK 4

ELA
POETRY

OVERVIEW:

This week you will practice reading poetry and answering questions.

Poetry can be challenging, as it often employs figurative language, symbolism, and multiple layers of meaning. However, with the right strategies and practice, you can improve your ability to analyze and appreciate poems.

Strategies:

1. Read actively: Engage with the poem by asking questions, making predictions, and annotating the text as you read.

2. Identify the theme: Look for the central message or underlying meaning of the poem. Consider how the poet explores universal human experiences or emotions.

3. Analyze the structure: Examine how the poem is organized, including its stanza structure, rhyme scheme, and meter. Consider how these elements contribute to the poem's meaning.

4. Explore figurative language: Identify and interpret metaphors, similes, personification, and other literary devices. Think about how these devices enhance the poem's impact and meaning.

5. Read aloud: Reading a poem aloud can help you better understand its rhythm, tone, and emotional content.

DAY 1
WEEK 4

ELA
POETRY

Directions: Read this poem, then answer the questions.

Entropy's Embrace

In the crevices of time's inexorable march,
Where chaos and order intertwine,
Lies the essence of entropy's arch,
A dance of energy, a paradigm divine.

From the cosmic dawn's incipient light,
To the twilight of stars' dying embers,
Entropy weaves its tapestry, day and night,
A universal truth that forever remembers.

In the microcosmic realm of human affairs,
Entropy's hand guides the ebb and flow,
From birth to death, from laughter to tears,
A constant reminder of the path we undergo.

Yet in the face of entropy's unyielding reign,
We strive to create, to build, to inspire,
A fleeting defiance, a tenuous refrain,
Against the inevitable, a flickering fire.

For in the end, as the universe grows cold,
And all that was, surrenders to entropy's embrace,
We shall find solace in the stories told,
Of the beauty and strife of the human race.

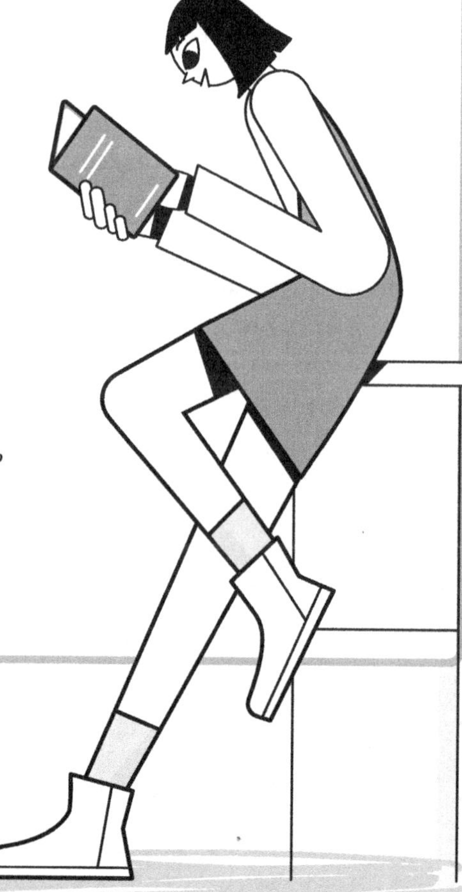

1. The poem personifies entropy as:

   A. A benevolent guide
   B. A malevolent force
   C. An indifferent observer
   D. A divine choreographer

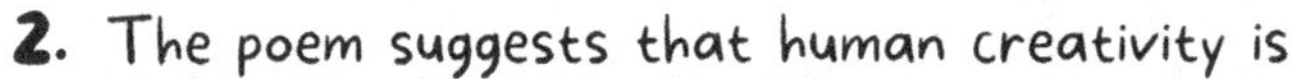

2. The poem suggests that human creativity is:

   A. A futile endeavor in the face of entropy
   B. A temporary defiance against the inevitable
   C. A product of entropy's influence
   D. A means to transcend entropy's grasp

3. The final stanza implies that:

   A. Entropy is a force to be feared and resisted
   B. The human race is ultimately insignificant
   C. The stories of humanity will outlast the universe itself
   D. There is beauty in the acceptance of entropy's embrace

4. The poem's tone can be best described as:

   A. Hopeful and optimistic
   B. Angry and defiant
   C. Resigned and contemplative
   D. Sorrowful and despairing

Let's get some fitness in! Go to page 169 to try some fitness activities.

DAY 2 WEEK 4

ELA POETRY

Directions: Read this poem, then answer the questions.

Whispers in the Wind

Amid the twilight's velvet grace,
Echoes stir the ancient trees,
Whispers cloak the night's embrace,
Secrets carried in the breeze.

Each leaf, a whisper of the past,
Tales spun from a thousand threads,
Moments caught in amber cast,
Silent songs where time treads.

Beneath the moon's watchful eyes,
Shadows dance to silent tunes,
Winds weave truth through cloaked lies,
Stories told beneath the moons.

What secrets do the soft winds hide?
What tales of old do they confide?
As stars peer down with gentle pride,
In the quiet, our thoughts collide.

1. Which literary device is predominantly used in the line "Echoes stir the ancient trees"?

   A. Metaphor
   B. Alliteration
   C. Personification
   D. Simile

2. What theme is primarily explored in the poem?

A. The excitement of urban life
B. The mystery and history of nature
C. The adventures of travel
D. The hardships of rural work

3. The repetition of the word 'whispers' throughout the poem emphasizes:

A. The loudness of the natural world.
B. The secretive and soft-spoken nature of the environment.
C. The aggressive communication between elements of nature.
D. The clarity and openness of the forest.

4. How does the poet use the moon in the poem?

A. As a symbol of loneliness and isolation.
B. As a source of light and guidance.
C. As a witness to the passing of time and stories.
D. As a disruptive force in nature.

Notes:

Directions: Read this poem, then answer the questions.

The Labyrinth of Identity

In the labyrinthine depths of the psyche's maze,
Where shadows dance and mirrors reflect,
The fragments of the self, a kaleidoscopic haze,
Converge and diverge, a journey to dissect.

The roles we play, the masks we wear,
Personas crafted for the world to see,
Yet beneath the façade, the truth laid bare,
A mosaic of contradictions, a complex tapestry.

Are we the sum of our memories, the tales we tell?
The experiences that shape our core within?
Or are we the aspirations that in our hearts dwell,
The dreams that drive us, the goals we yearn to win?

Nature and nurture, the eternal debate,
The influence of genes, the impact of our peers,
In the crucible of life, a delicate balance, a fragile state,
Forging the essence of our character, our hopes and fears.

The search for meaning, a perennial quest,
In the labyrinth of identity, a path to find,
A journey of self-discovery, a lifelong test,
To unravel the enigma, the puzzle of the mind.

For in the end, the truth we must embrace,
Is that the self is a mutable, ever-changing form,
A canvas painted with hues of time and space,
A masterpiece, forever evolving, forever reborn.

5. The labyrinth in the poem serves as a metaphor for:

   A. The depths of the human mind and identity
   B. The intricacies of human relationships
   C. The complexity of the world around us
   D. The challenges of navigating societal expectations

6. The phrase "kaleidoscopic haze" in the first stanza suggests that:

   A. Identity is a clear and easily defined concept
   B. The self is composed of many shifting and interconnected elements
   C. Memories are unreliable and constantly changing
   D. The psyche is a place of confusion and danger

7. The phrase "perennial quest" in the fifth stanza implies that the search for meaning is:

   A. A unique experience for each individual
   B. A futile and meaningless endeavor
   C. An ongoing and recurring journey throughout life
   D. A goal that is easily achieved and maintained

Notes:

Let's get some fitness in! Go to page 169 to try some fitness activities.

OVERVIEW:

In this week, you will learn about **sine, cosine**, and **tangent**, often referred to collectively as the "trigonometric ratios". Let's define the three terms.

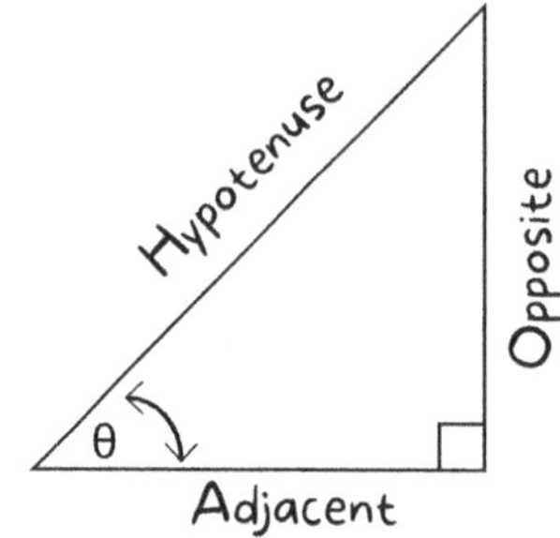

Right Traingle Diagram

1. Sine (sin):

For a given angle θ in a right triangle:

Sine is the ratio of the length of the side that is **opposite** this angle to the length of the **hypotenuse** (the side opposite the right angle).

Mathematically,

$$\sin(\theta) = \frac{\text{opposite}}{\text{hypotenuse}}$$

2. Cosine (cos):

For a given angle θ in a right triangle:

Cosine is the ratio of the length of the side that is **adjacent** to this angle (i.e., the side next to the angle but not the hypotenuse) to the length of the **hypotenuse**.

Mathematically,

$$\cos(\theta) = \frac{\text{adjacent}}{\text{hypotenuse}}$$

3. Tangent (tan):

For a given angle θ in a right triangle:

Tangent is the ratio of the **sine** of the angle to the **cosine** of the angle. It can also be understood as the ratio of the side opposite the angle to the side adjacent to it.

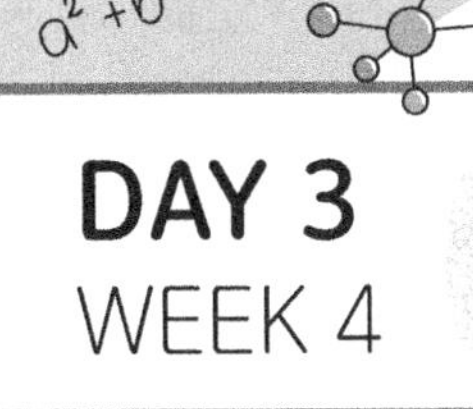

DAY 3
WEEK 4

MATH
TRIG FUNCTIONS

Mathematically,

$$\tan(\theta) = \frac{\sin(\theta)}{\cos(\theta)} = \frac{\text{opposite}}{\text{adjacent}}$$

You **must** remember the sin cos tan formulas for trigonometry. Fortunately, there is a great acronym to remember this.

* Remember the mnemonic word **SOHCAHTOA**.

Another mnenomic that you may prefer is to remember the phrase "Studying Our Homework Can Always Help To Obtain Achievement". We recommend remembering the word **SOHCAHTOA**.

SOH ⟶ $\sin(\theta) = \frac{\text{Opposite}}{\text{Hypotenuse}}$

CAH ⟶ $\cos(\theta) = \frac{\text{Adjacent}}{\text{Hypotenuse}}$

TOA ⟶ $\tan(\theta) = \frac{\text{Opposite}}{\text{Adjacent}}$

There are **three** other trig functions we must know, **cosecant (csc)**, **secant (sec)**, and **cotangent (cot)**. These are the **reciprocolas** of the sin, cos, and tan functions. So it's actually pretty easy to remember!

*Important Note: Many students confuse and think cosecant is the reciprocal of sin and secant to be the reciprocal of cos, which is **NOT** true. Do not confuse these two up.

- Cosecant (csc) is the reciprocal of sin
- Secant (sec) is the reciprocal of cos
- Cotangent (cot) is the reciprocal of tan

$$\csc(\theta) = \frac{\text{hypotenuse}}{\text{opposite}}, \quad \sec(\theta) = \frac{\text{hypotenuse}}{\text{adjacent}}, \quad \cot(\theta) = \frac{\text{adjacent}}{\text{opposite}}$$

DAY 3 WEEK 4

MATH

TRIG FUNCTIONS

Let's take a look at a few practice exercises.

Example 1:

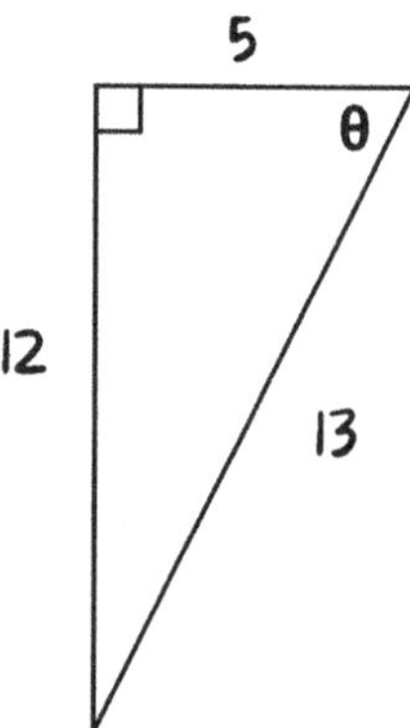

What is $\cos \theta$?

We have a right-sided triangle here with all 3 sides labeled with their lengths.

We know $\cos\theta = \dfrac{\text{adjacent}}{\text{hypotenuse}}$

Looking at **θ**, we can see **5** is adjacent and 13 is the hypotenuse.

So $\cos\theta = \dfrac{5}{13}$

Example 2:

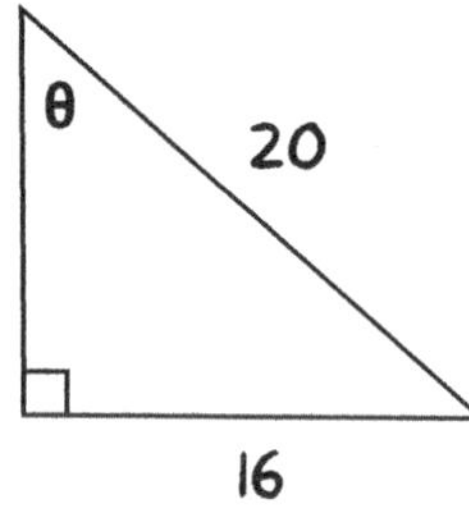

What is $\cos \theta$?

We have a right-sided triangle here with **2 sides** labeled.

We know $\cos\theta = \dfrac{\text{adjacent}}{\text{hypotenuse}}$

We do **not** know the adjacent value, because it was not provided in the diagram. How can we solve for this missing value?

Pythagorean Theorem!

We have $a^2 + 16^2 = 20^2$

$a^2 + 256 = 400$

$a^2 = 144$ (Now take the square root of both sides to get $a = 12$).

$a = 12$

Let's go ahead and relabel our right triangle.

Now we know the adjacent value is 12 and the hypotenuse is 20.

DAY 3 WEEK 4

MATH
TRIG FUNCTIONS

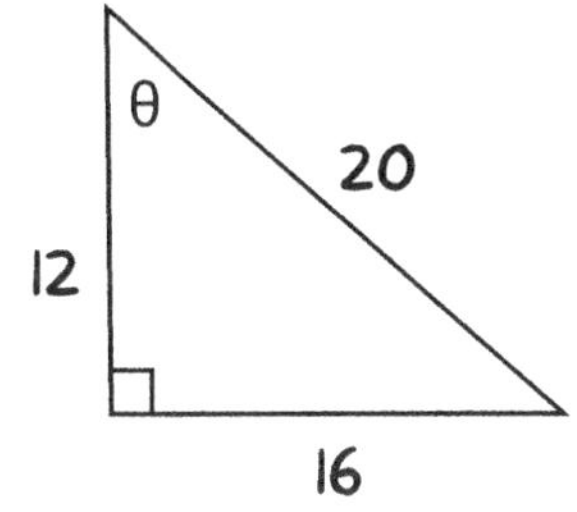

So, $\cos \theta = \frac{12}{20}$. We can simplify this fraction by dividing the numerator and denominator by 4 to get $\frac{3}{5}$.

So $\cos \theta = \frac{3}{5}$.

Example 3:

What is $\cot \theta$?

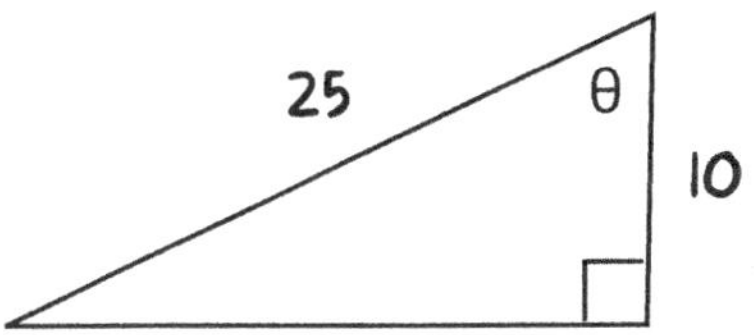

We have a right-sided triangle here with **2 sides** labeled.

We know $\cot \theta = \frac{\text{adjacent}}{\text{opposite}}$

We know the adjacent value, but we **do not** know the opposite value.

How can we solve for this missing value?

Pythagorean Theorem!

We have $10^2 + b^2 = (25)^2$

$100 + b^2 = 625$

$b^2 = 525$ (Now take the square root of both sides to get $\sqrt{525}$.

We can simplify $\sqrt{525}$ to $\sqrt{25} \times \sqrt{21} \rightarrow 5\sqrt{21}$

Let's go ahead and relabel our right triangle.

Now we know the adjacent value is 10 and the opposite value is $5\sqrt{21}$.

So $\cot \theta = \frac{10}{5\sqrt{21}}$. This is an interesting situation!

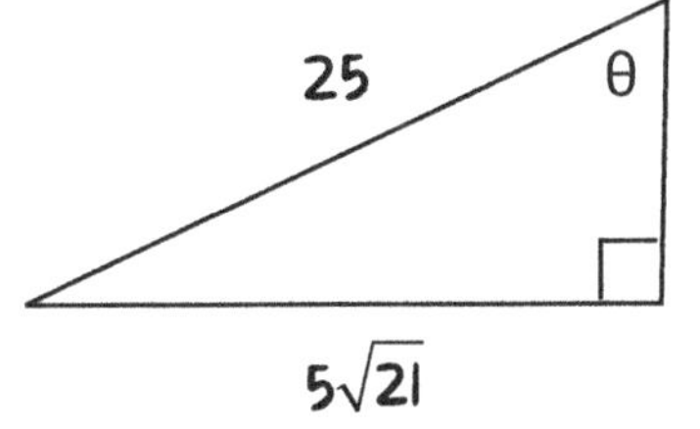

Whenever there is a radical in the denominator of a fraction, it must be rationalized. You should have covered this in math class last year before trigonometry. You can rationalize it by multiplying both the numerator and the denominator by that square root.

$$\frac{10}{5\sqrt{21}} \times \frac{5\sqrt{21}}{5\sqrt{21}} \times \frac{\overset{2}{\cancel{50}}\sqrt{21}}{\cancel{25} \times 21} = \frac{2\sqrt{21}}{21}$$

So $\cot \theta = \frac{2\sqrt{21}}{21}$

Your turn to practice trig functions!

DAY 3
WEEK 4

MATH
TRIG FUNCTIONS

Find the value of the trig function indicated. Use a separate piece of paper to show your work if needed. The triangles are not drawn to scale.

1. $\sec\theta$

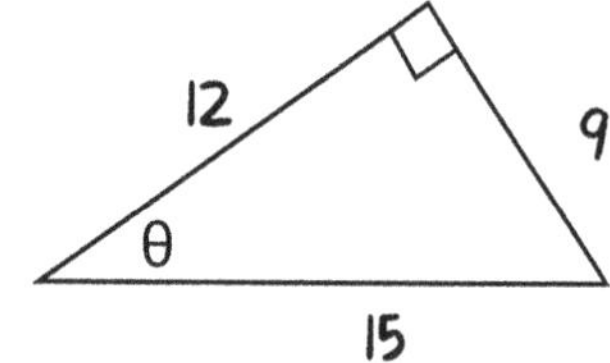

2. $\cos\theta$

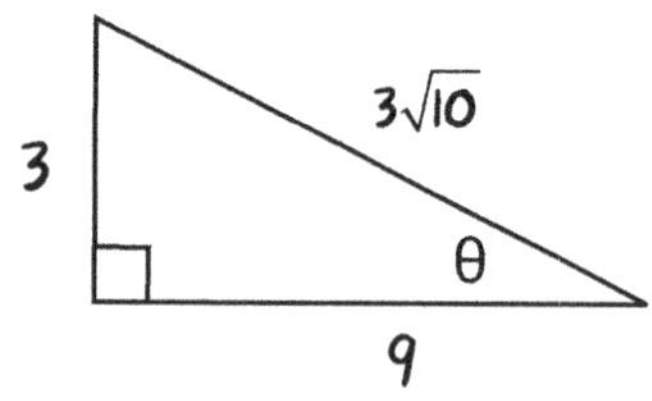

3. $\sec\theta$

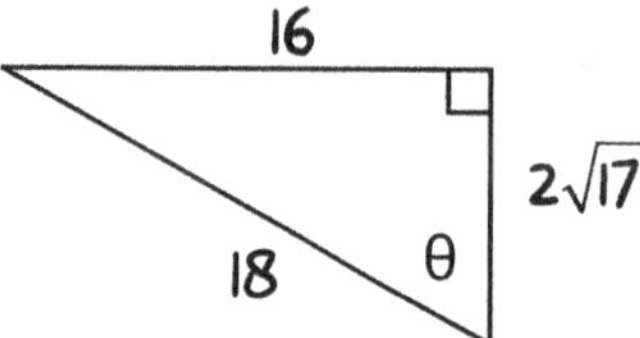

4. $\tan\theta$

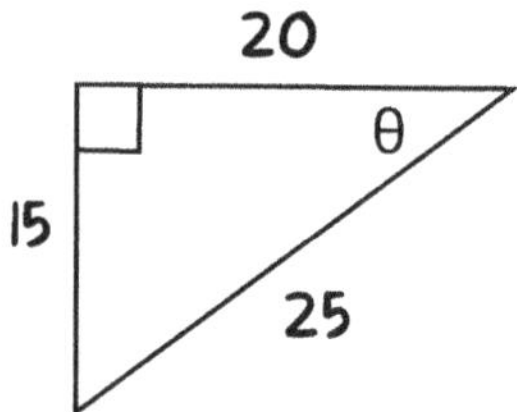

5. $\cot\theta$

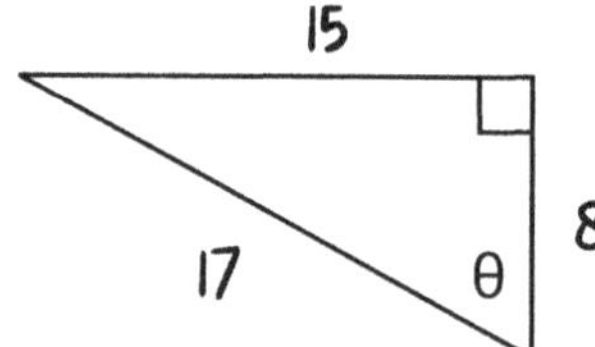

6. $\sin\theta$

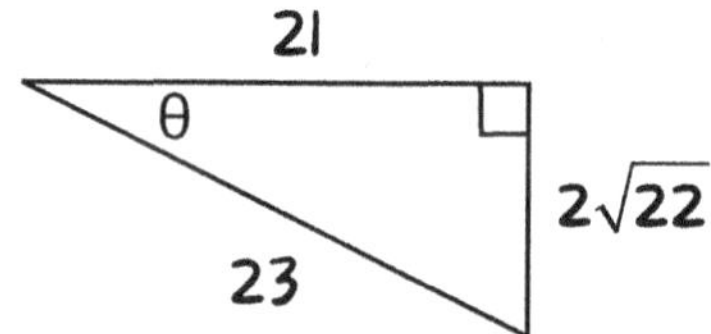

Let's get some fitness in! Go to page 169 to try some fitness activities.

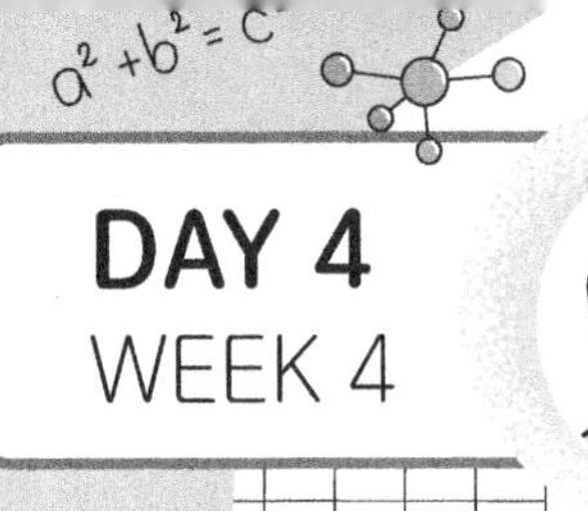

DAY 4
WEEK 4

MATH
TRIG FUNCTIONS

Find the value of the trig function indicated. Use a separate piece of paper to show your work if needed. The triangles are not drawn to scale.

1. $\tan\theta$

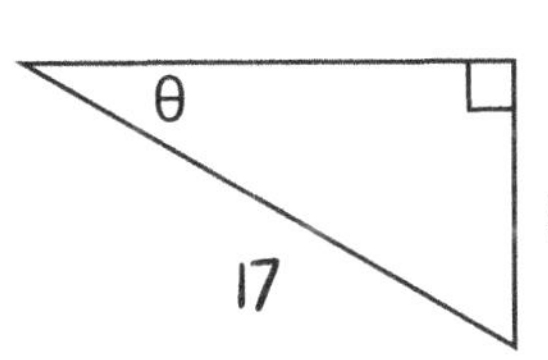

2. $\csc\theta$

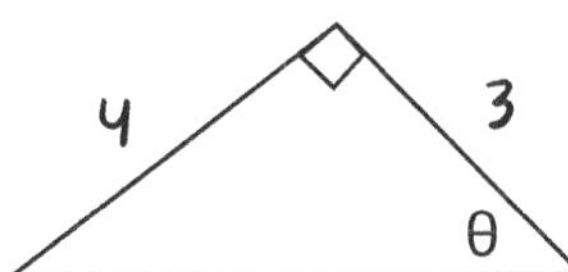

3. $\cos\theta$

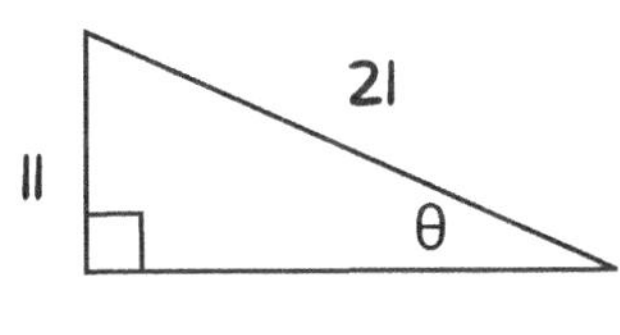

4. $\sin\theta$

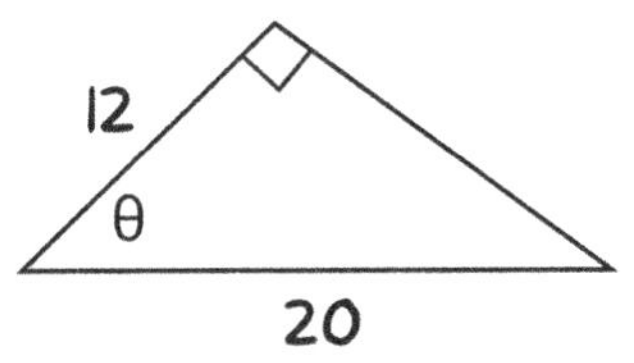

5. $\sec\theta$

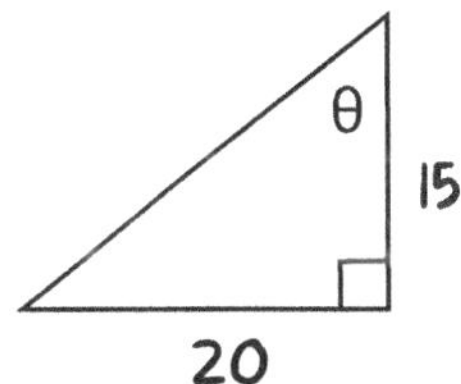

6. $\cos\theta$

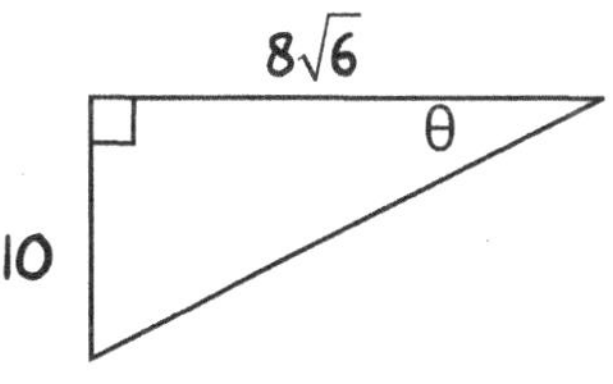

7. $\sec\theta$

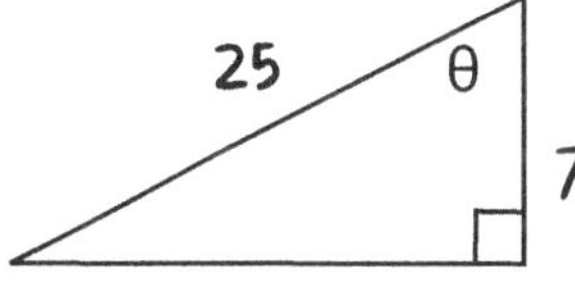

8. $\cot\theta$

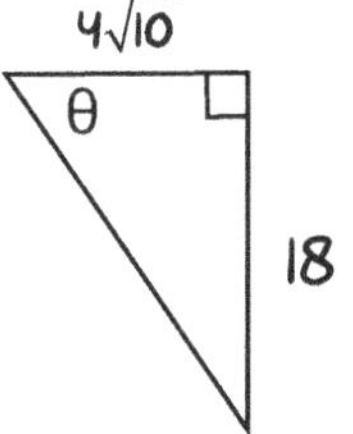

DAY 4
WEEK 4

MATH
TRIG FUNCTIONS

9. $\sin\theta$

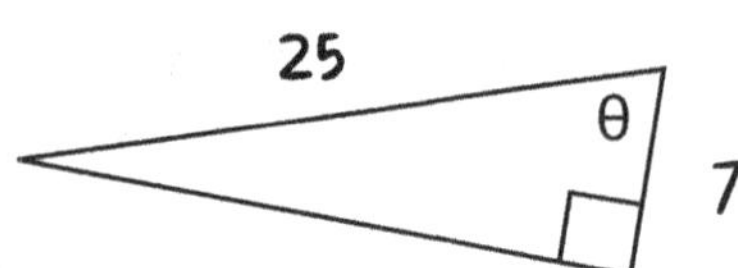

10. $\sec\theta$

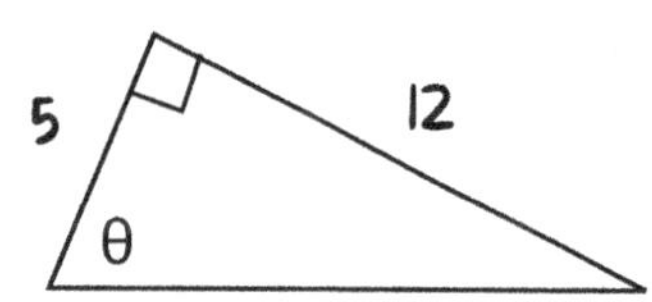

11. $\cot\theta$

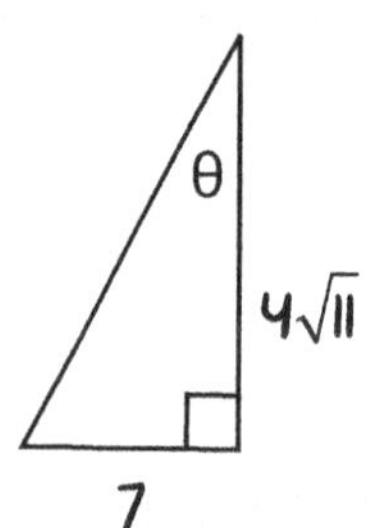

12. $\sec\theta$

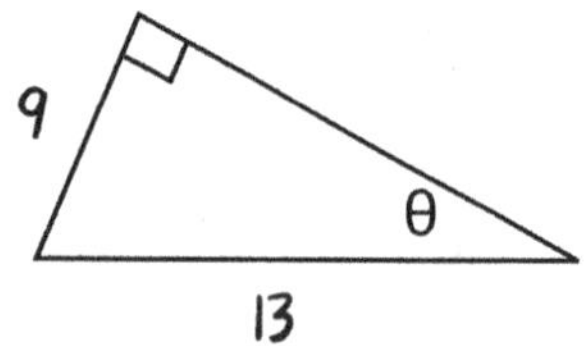

13. $\csc\theta$

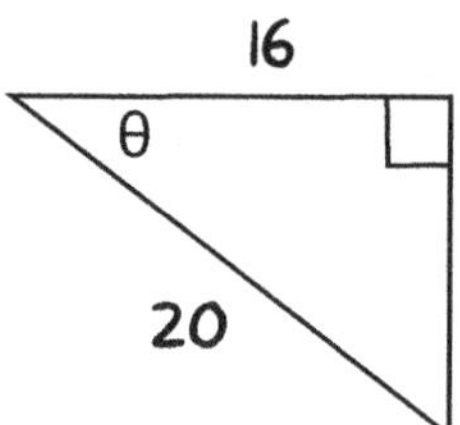

14. $\csc\theta$

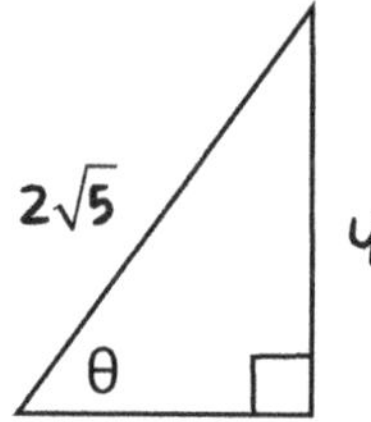

15. $\sin\theta$

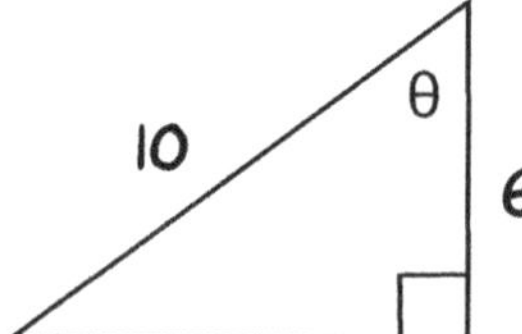

Let's get some fitness in! Go to page 169 to try some fitness activities.

DAY 5
WEEK 4

MATH
TRIG FUNCTIONS

Find the value of the trig function indicated. Use a separate piece of paper to show your work if needed. The triangles are not drawn to scale.

1. $\cot\theta$

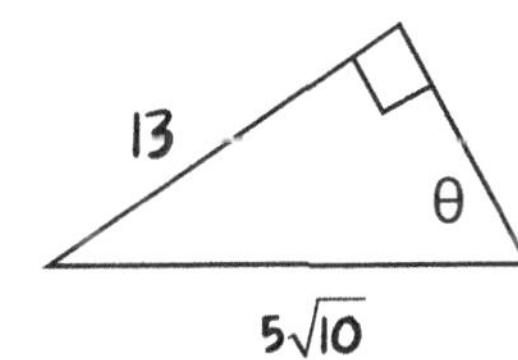

2. $\csc\theta$

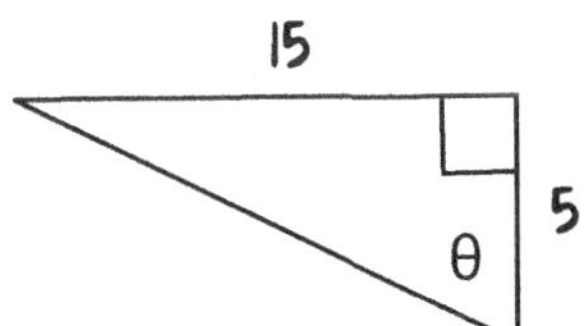

3. $\sin\theta$

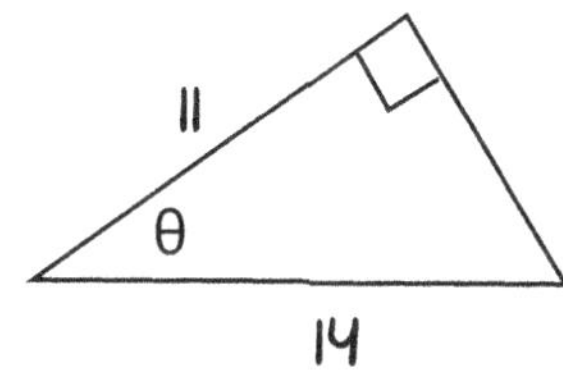

4. $\cos\theta$

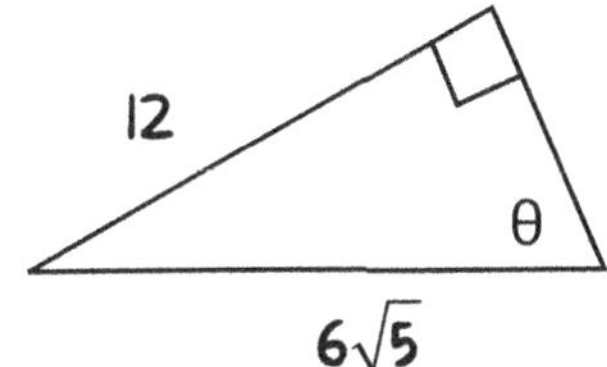

5. $\tan\theta$

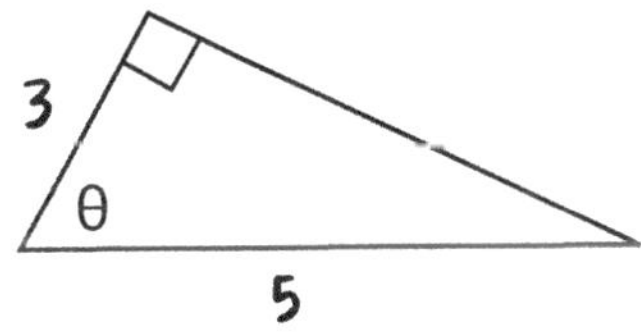

6. $\cos\theta$

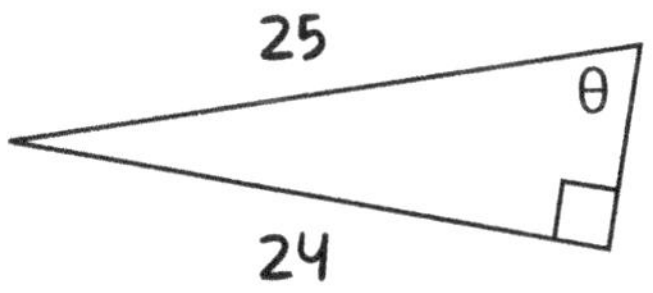

7. $\cot\theta$

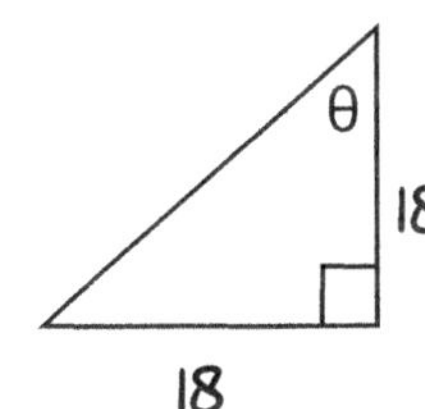

8. $\tan\theta$

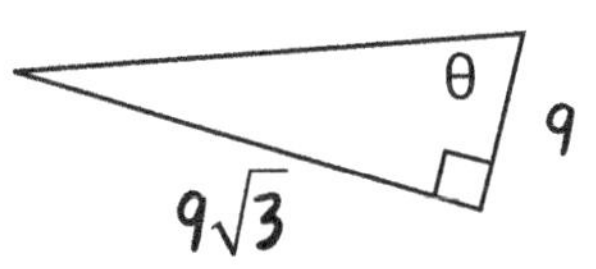

DAY 5 WEEK 4

MATH

TRIG FUNCTIONS

9. $\cos\theta$

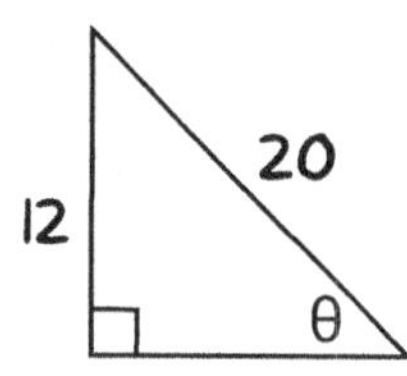

13. $\sin\theta$

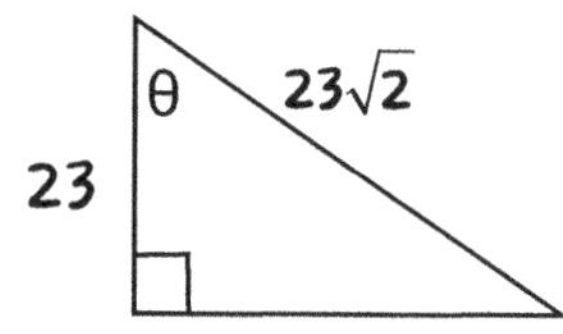

10. $\sec\theta$

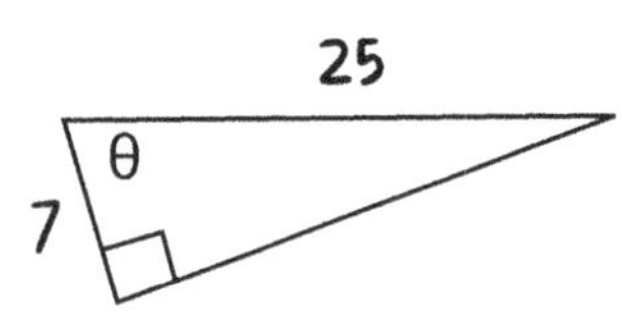

14. $\sec\theta$

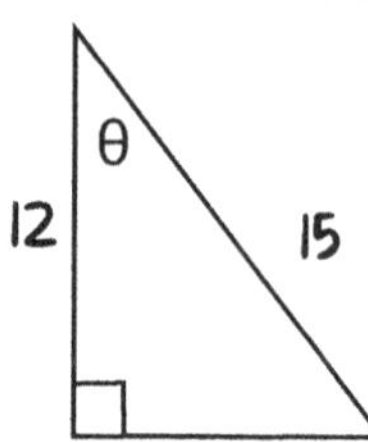

11. $\tan\theta$

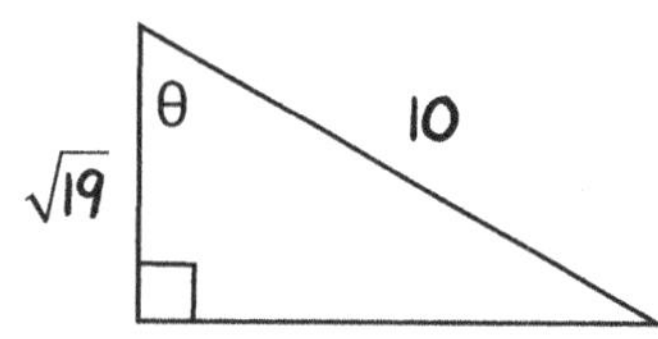

15. $\sin\theta$

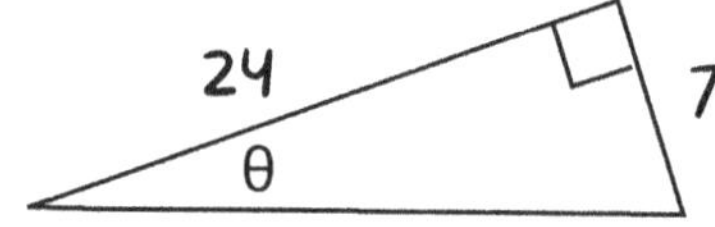

12. $\csc\theta$

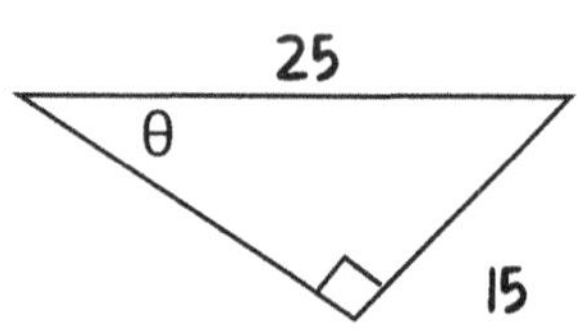

16. $\csc\theta$

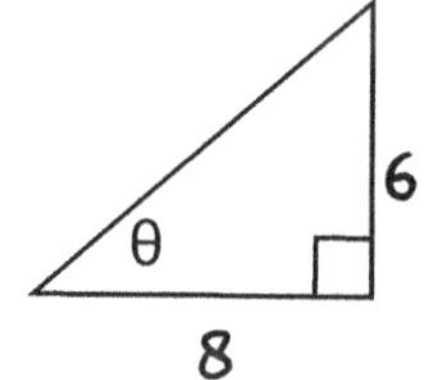

Let's get some fitness in! Go to page 169 to try some fitness activities.

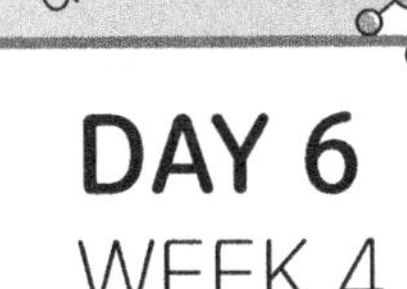

DAY 6
WEEK 4

ELA
VOCABULARY

Directions: Read each sentence carefully and use context clues to determine the meaning of the underlined word. Choose the best definition for the underlined word from the options provided. Then, write a sentence using the word in a new context, demonstrating your understanding of its meaning and usage. You can use the Internet to look up the exact definition after answering the questions.

1. The new laws enacted by the parliament engendered considerable **consternation** among the population, who felt unprepared for such sweeping changes.

A. confusion
B. excitement
C. contentment
D. anger

Sentence:

....

....

....

2. In the debate, the candidate's **reticence** about her political strategies only fueled more speculation and distrust among the voters.

A. openness
B. reluctance
C. eagerness
D. corruptness

Sentence:

....

....

....

3. Despite the CEO's reassurances, the **obfuscation** of financial details in the report led to increased skepticism from the investors.

A. clarification
B. transparency
C. simplification
D. complication

Sentence:

4. During the trial, the witness's testimony was **equivocal**, leaving the jury perplexed about the true sequence of events.

A. ambiguous and unclear
B. clear and straightforward
C. detailed and thorough
D. false and convincing

Sentence:

5. Her argument was **trenchant**, making a clear and effective point that was hard to contest.

A. weak and unconvincing
B. powerful and compelling
C. polite and respectful
D. vague and general

Sentence:

Let's get some fitness in! Go to page 169 to try some fitness activities.

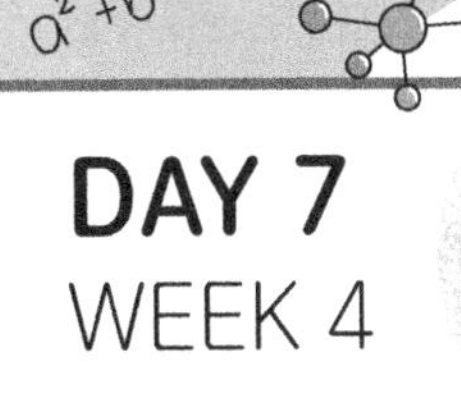

DAY 7
WEEK 4

SAT
ELA PREP

Directions: Work through these questions carefully.

1. In their rulings, judges often reference social science research to support their decisions, particularly in cases involving human behavior and societal norms. Judge Clara Thompson suggests that while judges tend to favor research that corroborates their own viewpoints, the most persuasive rulings critically engage with and challenge contrary findings; thus, incorporating diverse research could ________

Which choice most logically completes the text?

A. permit judges to disregard less relevant social science studies in their rulings.

B. enhance the credibility and depth of the arguments presented in judicial opinions.

C. simplify judicial opinions for a broader audience unfamiliar with social sciences.

D. align judicial opinions more closely with mainstream social science perspectives.

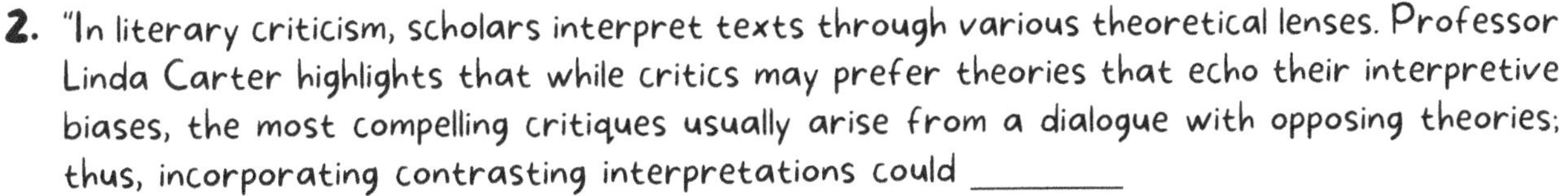

2. "In literary criticism, scholars interpret texts through various theoretical lenses. Professor Linda Carter highlights that while critics may prefer theories that echo their interpretive biases, the most compelling critiques usually arise from a dialogue with opposing theories; thus, incorporating contrasting interpretations could ________

Which choice most logically completes the text?

A. allow literary critics to avoid engaging with the text itself.

B. ensure uniformity in critical responses to literary works.

C. limit the scope of interpretations to those most widely accepted.

D. enhance the richness and persuasiveness of literary analyses.

3. Nutritionists at the local health clinic have developed a program to educate individuals about healthy eating habits. The program's success has grown after asking participants if ________ understood the information and could apply it in daily life.

Which choice completes the text so that it conforms to the conventions of Standard English?

A. one

B. he

C. they

D. it

4. Recent research suggests that the mimic octopus (Thaumoctopus mimicus) has the unique ability to alter its color and texture to blend in with its surroundings, effectively camouflaging ________ from predators.

Which choice completes the text so that it conforms to the conventions of Standard English?

A. itself

B. themselves

C. it

D. him

5. Researchers at the Galapagos Marine Institute are currently studying the diet of the blue-footed booby to assess whether environmental pollutants have ________ impact on their reproductive success. The goal is to determine if exposure to chemicals correlates with changes in birth rates.

Which choice completes the text with the most logical and precise word or phrase?

A. a detrimental

B. an undetectable

C. a negligible

D. an advantageous

Notes:

Let's get some fitness in! Go to page 169 to try some fitness activities.

WEEK 5

GRADE 10-11

This week, you will read passages and answer the questions for the ELA section. For math, you will practice solving for missing sides and angles using the trig functions you learned in the previous week.

Directions: Read the passage below. Then answer the questions.

The Paradox of Progress

In an era defined by rapid technological escalation, the phenomenon of automation has burgeoned as a pivotal factor in reshaping socio-economic landscapes across the globe. This relentless pursuit of efficiency through mechanization not only promises unprecedented productivity but also engenders profound ethical dilemmas and societal shifts.

The displacement of labor, a conspicuous consequence of automation, epitomizes the paradox inherent in technological advancement. As robotic systems and artificial intelligence infiltrate industries, the fabric of traditional employment is irrevocably altered. This metamorphosis extends beyond mere job loss; it heralds a fundamental transformation in the nature of work itself. Where once manual competence and routine cognitive functions dominated, now adaptability, creativity, and technological fluency command premium significance.

Simultaneously, the ascendancy of algorithms in decision-making processes magnifies latent biases within societal structures. These automated systems, designed to emulate optimal decision-making, often derive their logic from historical data fraught with entrenched prejudices. Consequently, without vigilant oversight, they inadvertently perpetuate and amplify these biases, affecting everything from employment opportunities to judicial sentencing.

Moreover, the environmental ramifications of automation present a complex juxtaposition of potential ecological benefits against severe detriments. While more efficient processes may reduce resource wastage, the lifecycle of technologically sophisticated equipment frequently incurs substantial ecological debt, exacerbated by the non-biodegradable waste and energy-intensive production practices associated with modern technologies.

As we navigate this labyrinth of technological evolution, the ethical implications of automation challenge us to reevaluate our societal priorities and the trajectory of human advancement. It compels us to ask: In our quest for progress, what are we willing to sacrifice, and to what extent can we mitigate the inherent trade-offs of this progress?

DAY 1 WEEK 5

ELA
READING COMPREHENSION PASSAGE

1. What does the term "**epitomizes**" most closely mean as used in the passage?

   A. Represents
   B. Eliminates
   C. Minimizes
   D. Overstates

2. Which of the following best captures the inference about the future of employment from the passage?

   A. Future employment will primarily require manual skills and routine cognitive abilities.
   B. Employment opportunities will expand universally due to increased productivity from automation.
   C. The nature of employment will shift to prioritize adaptability and technological proficiency.
   D. Job opportunities will remain stable as industries adapt to technological advancements.

3. Based on the passage, how do algorithms potentially exacerbate social inequalities?

   A. By optimizing decision-making processes to be more just and unbiased.
   B. Through replicating and magnifying existing societal biases present in their training data.
   C. By reducing the overall efficiency of industries they are implemented in.
   D. Algorithms have no significant impact on social dynamics according to the passage.

4. The author refers to automation's environmental impact as a "complex juxtaposition." What can be inferred from this description?

   A. Automation unequivocally benefits the environment by reducing waste.
   B. The environmental effects of automation are straightforward and predominantly positive.
   C. There are significant trade-offs between potential ecological benefits and detriments in automation.
   D. Automation has no real impact on the environment.

Let's get some fitness in! Go to page **169** to try some fitness activities.

Directions: Read the passage below. Then answer the questions.

The Enigma of Consciousness

Consciousness, the ineffable essence of subjective experience, remains one of the most perplexing and recondite enigmas in the realm of cognitive science. This nebulous phenomenon, which endows us with an intimate awareness of our thoughts, emotions, and perceptions, has long confounded philosophers and scientists alike. The ontological nature of consciousness eludes precise definition, as it transcends the reductionist paradigms of conventional scientific inquiry.

At the crux of the conundrum lies the seemingly insurmountable chasm between the objective, third-person perspective of neuroscience and the subjective, first-person experience of conscious awareness. While advanced neuroimaging techniques have illuminated the intricate workings of the brain, the neural correlates of consciousness remain shrouded in mystery. The activation of specific cerebral regions, though correlated with conscious states, fails to adequately explicate the qualitative character of subjective experience.

Moreover, the question of qualia, the ineffable, intrinsic properties of conscious experiences, further compounds the enigma. The subjective sensation of the redness of red, the pungency of a rose's fragrance, or the melancholic strains of a violin cannot be reduced to mere neural activity. This explanatory gap between the physical substrate of the brain and the phenomenal nature of consciousness has led some philosophers to propose that consciousness may be an fundamental, irreducible aspect of reality itself.

The hard problem of consciousness, as articulated by the philosopher David Chalmers, encapsulates the difficulty in explaining how subjective experiences can arise from objective, physical processes. While the "easy problems" of consciousness, such as the neural mechanisms underlying attention, memory, and perception, are amenable to scientific investigation, the hard problem remains obstinately unresolved. Some theorists posit that a comprehensive understanding of consciousness may require a radical revision of our current scientific paradigms, potentially invoking exotic concepts such as quantum entanglement or panpsychism.

Furthermore, the evolutionary origins of consciousness remain shrouded in uncertainty. While some scholars argue that consciousness emerged as an adaptive trait, conferring a selective advantage in complex social environments, others contend that it may be a spandrel, a byproduct of other evolutionary processes. The apparent continuity of consciousness across the animal kingdom, from the rudimentary awareness of insects to the rich inner lives of primates, adds another layer of complexity to the puzzle.

As we grapple with the profound implications of consciousness for our understanding of reality, free will, and the nature of the self, it becomes evident that unraveling this enigma will require an unprecedented synthesis of insights from neuroscience, philosophy, psychology, and potentially novel scientific paradigms. Until then, the mystery of consciousness will continue to captivate and inspire generations of thinkers, reminding us of the depths of the unknown that still pervade our understanding of the mind and its place in the cosmos.

1. The passage suggests that the "hard problem" of consciousness refers to the difficulty in:

   A. Understanding the neural mechanisms of attention and perception.

   B. Explaining the evolutionary origins of consciousness.

   C. Defining consciousness in a precise, scientific manner.

   D. Bridging the gap between physical brain processes and subjective experiences.

2. According to the passage, which of the following best describes the relationship between brain activity and conscious experiences?

   A. Specific patterns of neural activity fully explain the qualitative nature of consciousness.

   B. Neuroscience has successfully resolved the mystery of consciousness by mapping brain functions.

   C. While neural correlates of consciousness exist, they do not fully account for subjective experiences.

   D. Conscious experiences are entirely independent of brain activity and cannot be studied scientifically.

3. The passage mentions "qualia" as:

   A. A type of brain imaging technique used to study consciousness.

   B. The subjective, intrinsic properties of conscious experiences.

   C. A theory proposing that consciousness is a fundamental aspect of reality.

   D. The neural mechanisms underlying attention, memory, and perception.

4. Which of the following philosophical positions is NOT mentioned in the passage as a potential explanation for the nature of consciousness?

   A. Dualism

   B. Panpsychism

   C. Quantum entanglement

   D. Consciousness as an evolutionary byproduct

5. The author's tone in the passage can best be described as:

   A. Optimistic and certain about the scientific understanding of consciousness.

   B. Dismissive of the importance of studying consciousness.

   C. Deeply fascinated yet acknowledging the profound challenges in understanding consciousness.

   D. Highly critical of philosophical approaches to studying consciousness.

Let's get some fitness in! Go to page 169 to try some fitness activities.

DAY 3 WEEK 5

MATH

SOLVING FOR MISSING SIDES USING TRIG FUNCTIONS

You will need to use the calculator for these problems. Most schools and math classes have students use the TI-84 calculator. If you do not have this specific calculator, do not worry! Simply look your calculator model online and make sure you know how to set the mode to radian **or** degree. Here is an image of the TI-84 calculator below.

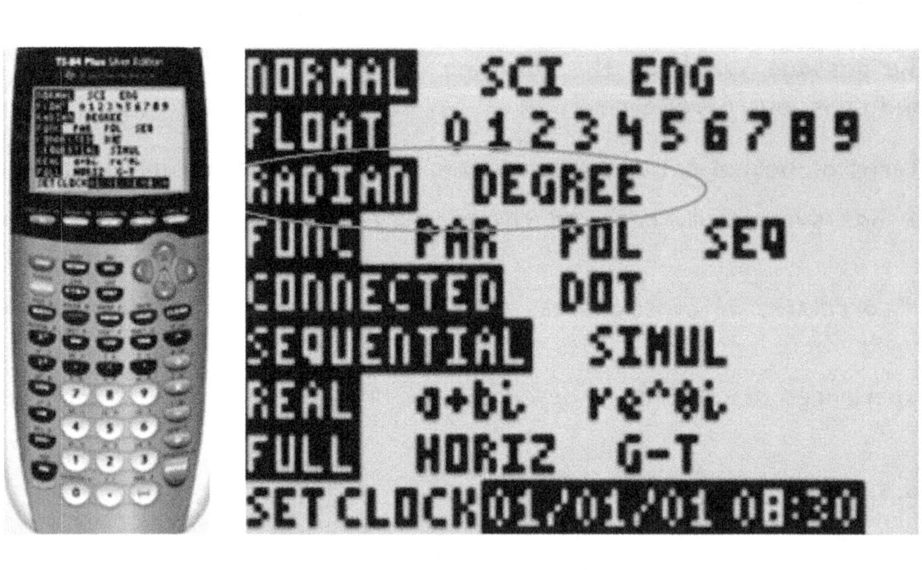

To toggle between Radian or Degree, you will need to click on the "2nd" button followed by the "Mode" button. Both of these buttons are found on the top left of the calculator. For these problems, you want to make sure you have the calculator set to **degrees**.

Example # 1

Carmen wants to measure the **height** of a tree. She walks exactly 100 feet from the **base** of the tree and looks up. The **angle** from the ground to the top of the tree is 33°. To the nearest foot, how tall is the tree?

We can use trigonometry to solve this problem. Notice that we can create a right-triangle.

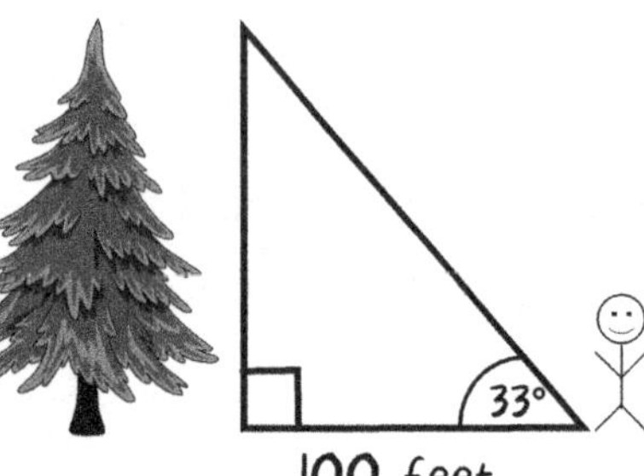

Note that 100 feet is **adjacent** to 33° and we need to find "x" which is **opposite** to 33°. Let's add in that new information to our diagram.

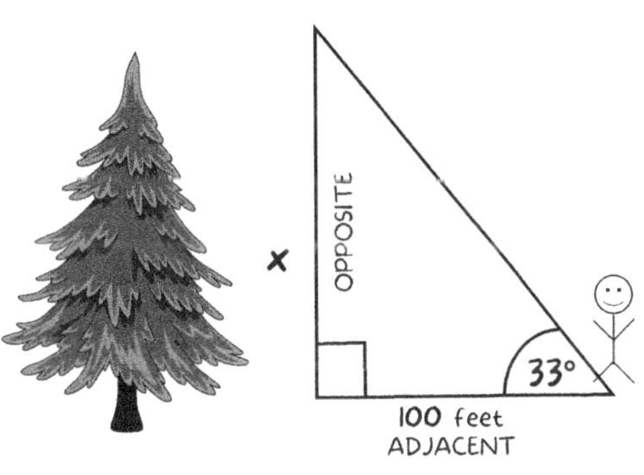

We have enough information to use trigonometry to solve this problem. We know the angel is 33°. We have the **adjacent** value and we need to find out the **opposite** value.

Which one of our trig function applies here? Remember the mnemonic device SOCAHTOA. **Here is a diagram reminder to refresh your memory.**

SOCAHTOA

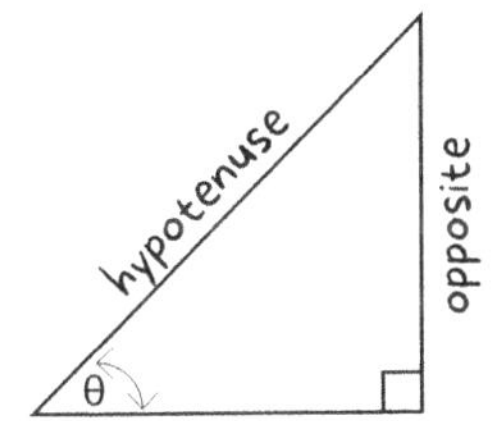

hypotenuse
θ
adjacent

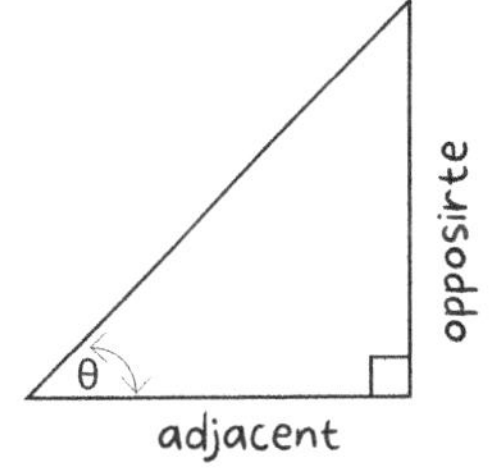

$\sin(\theta) = \frac{\text{opposite}}{\text{hypotenuse}}$ $\cos(\theta) = \frac{\text{adjacent}}{\text{hypotenuse}}$ $\tan(\theta) = \frac{\text{opposite}}{\text{adjacent}}$

SOH **CAH** **TOA**

We need to use $\tan \boldsymbol{\theta} = \frac{\text{opposite}}{\text{adjacent}}$. Plug in the known values.

$\tan 33° = \frac{x}{100}$.

$(100)\ (\tan 33°) = x$ **You need to use a calculator for all these problems.**

$64.94 = x$

The problem states to round up to the nearest foot, so the answer is **65 feet tall.**

Example # 2

Joseph is standing **31 metres** away from the base of a skyscraper. He looks up to the top of the building at a **78° angle**. How **tall** is the skyscraper? Round to the nearest meter.

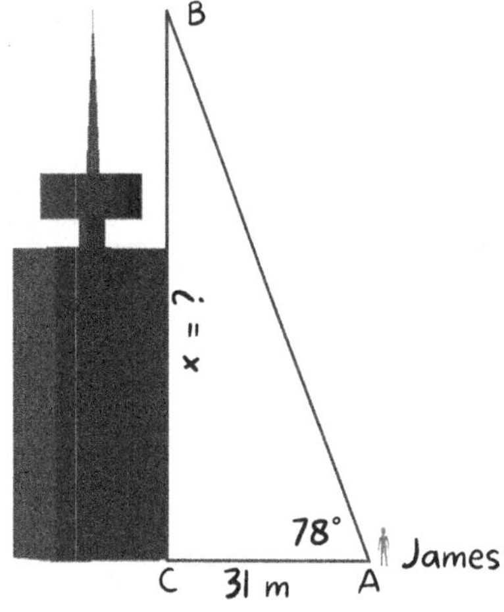

We need to use $\tan\theta = \frac{\text{opposite}}{\text{adjacent}}$. Plug in the known values.

$\tan 78° = \frac{x}{31}$

$4.7046 = \frac{x}{31}$

$145.8426 = x$

So $x = 145.8426$ meters.

The problem states to round up to the nearest meter, so the answer is **146 meters tall.**

Your turn to practice solving for missing sides using trig functions!

DAY 3
WEEK 5

MATH
SOLVING FOR MISSING SIDES USING TRIG FUNCTIONS

Directions: Find the measure of each side indicated. Round your answer to the nearest tenth. Be sure to include the units in your answer.

1.

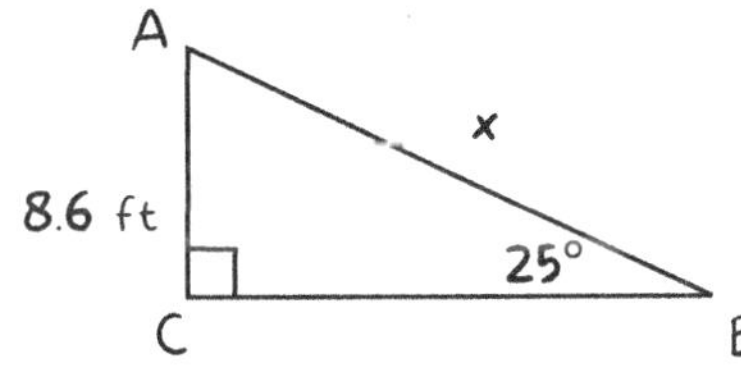

2.

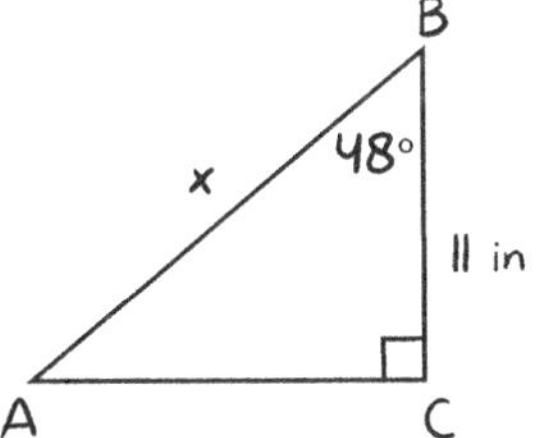

3.

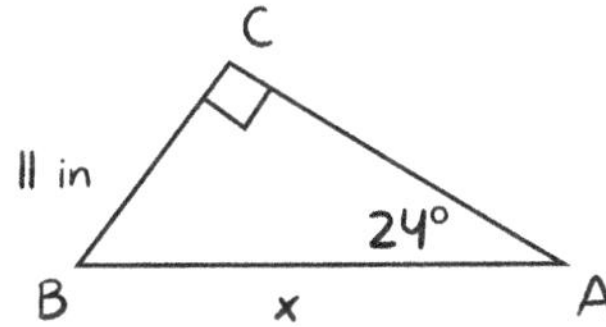

4.

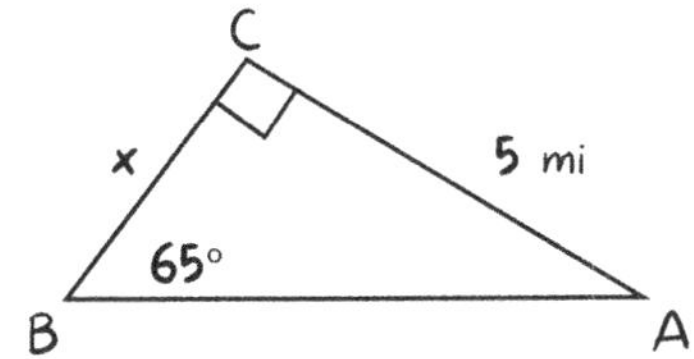

5.

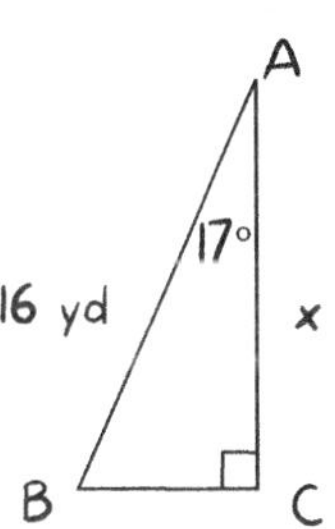

6.

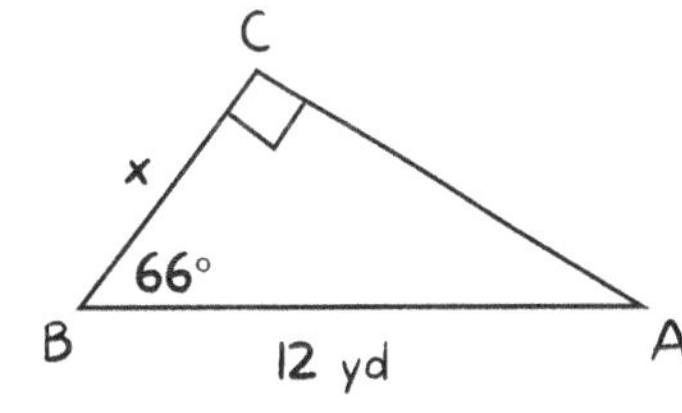

7.

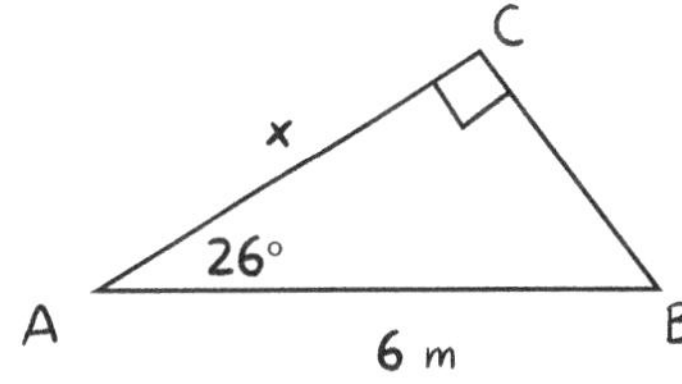

8.

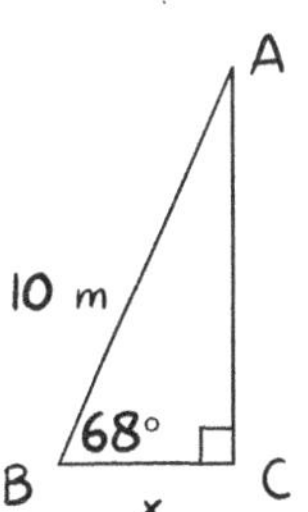

MATH

SOLVING FOR MISSING SIDES USING TRIG FUNCTIONS

9.

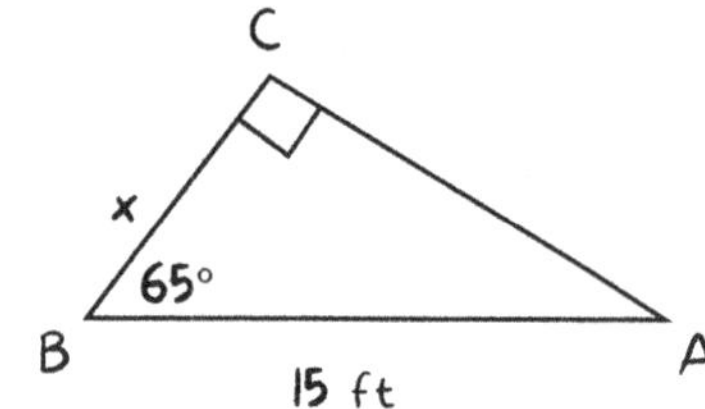

12.

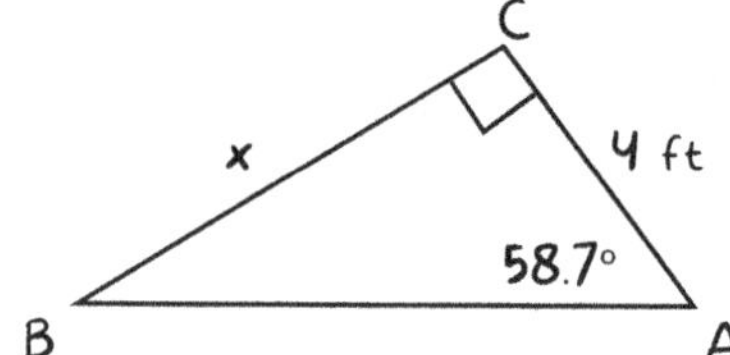

10.

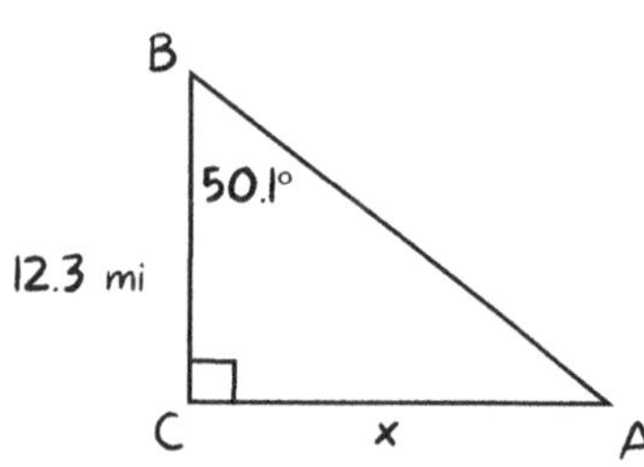

13.

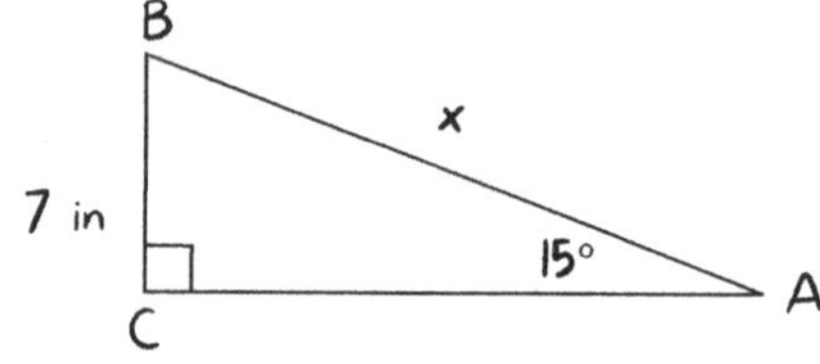

11.

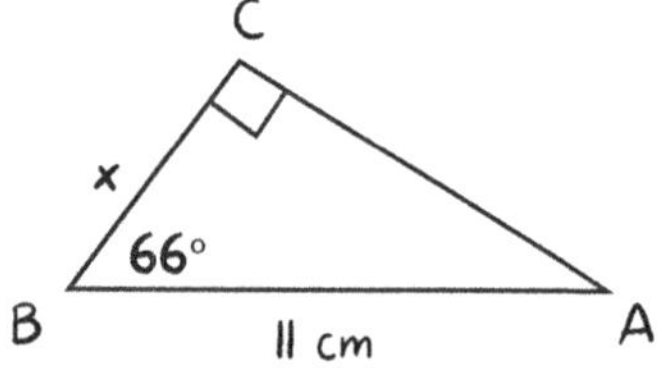

14.

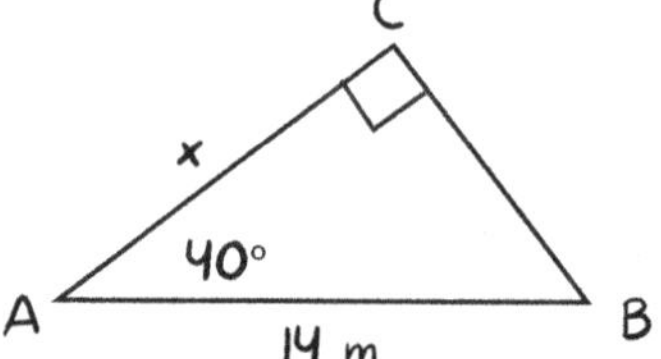

Let's get some fitness in! Go to page **169** to try some fitness activities.

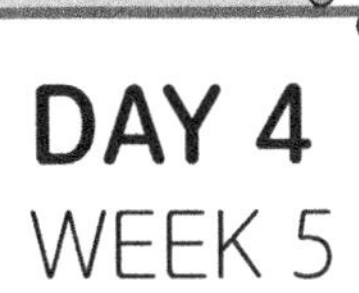

DAY 4 WEEK 5

MATH

SOLVING FOR MISSING SIDES USING TRIG FUNCTIONS

Directions: Find the measure of each side indicated. Round your answer to the nearest tenth. Be sure to include the units in your answer.

1.

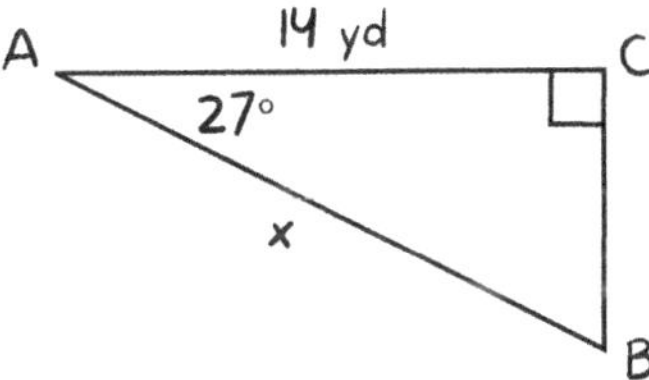

2.

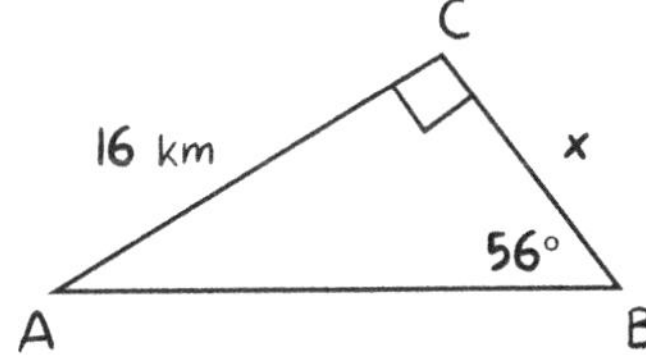

3.

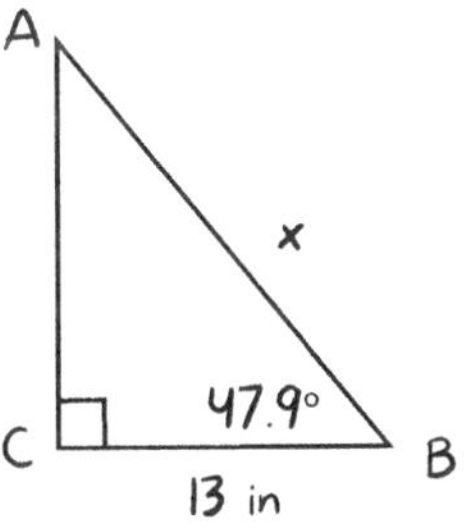

4.

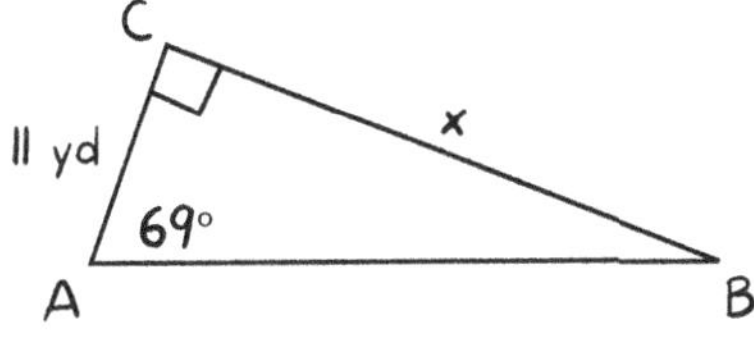

5.

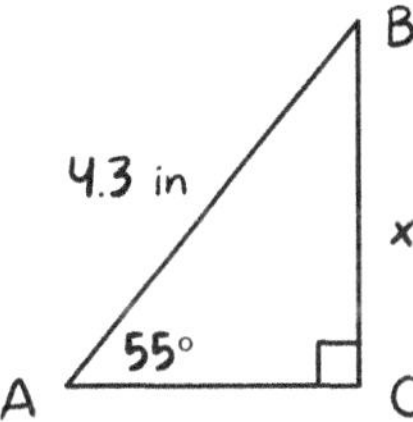

6.

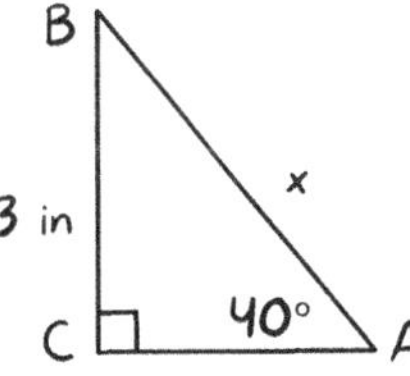

7.

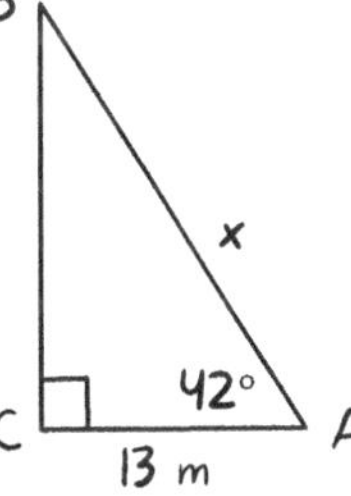

8.

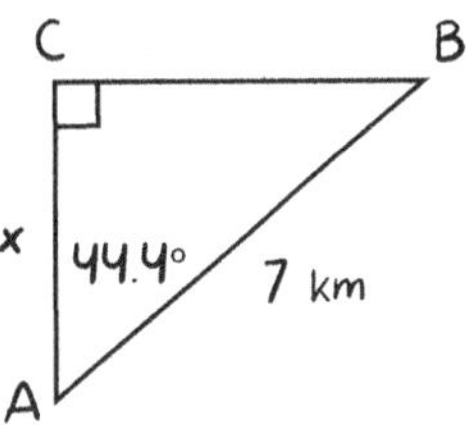

MATH

SOLVING FOR MISSING SIDES USING TRIG FUNCTIONS

9.

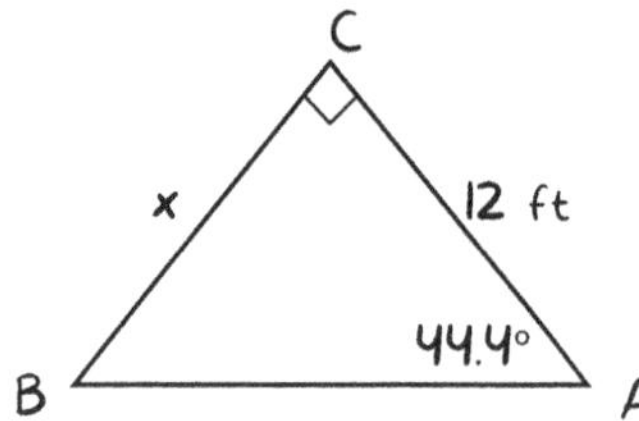

13.

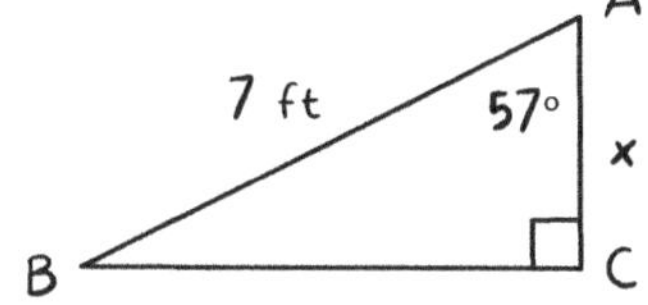

10.

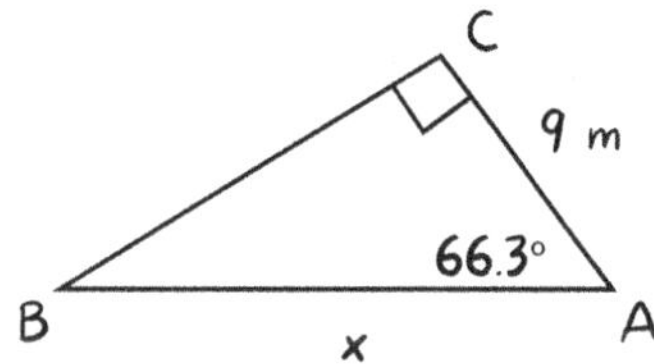

14.

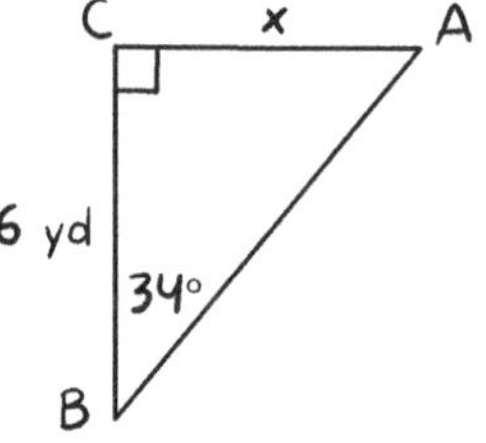

11.

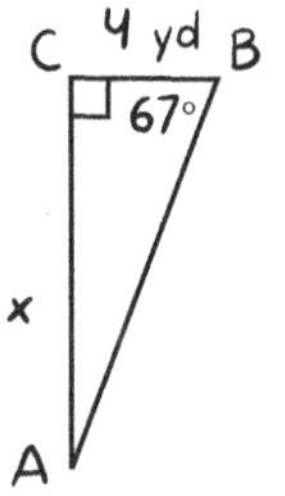

15.

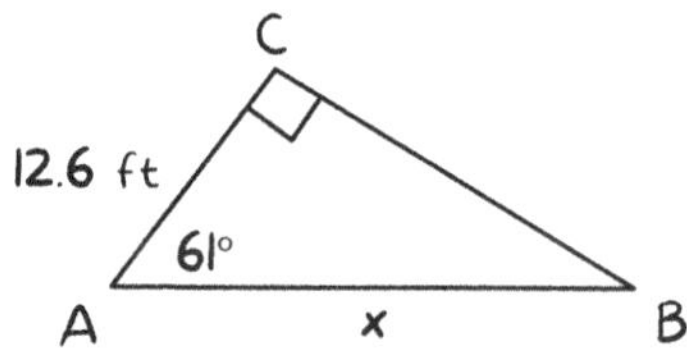

12.

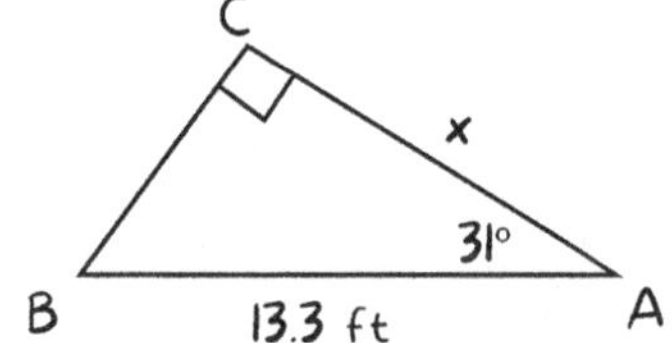

Let's get some fitness in! Go to page 169 to try some fitness activities.

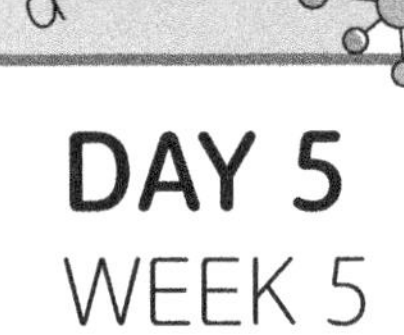

DAY 5 WEEK 5

MATH

SOLVING FOR MISSING ANGLES USING TRIG FUNCTIONS

Earlier we reviewed solving for missing slides using trig functions. Now, let's review some examples and solve for missing angles.

Example # 1

The ladder leans against a wall as shown. The foot of the ladder is **6 feet** from the wall. The ladder reaches a **height** of **15 feet** on the wall. What is the **angle** that the ladder makes with the wall?

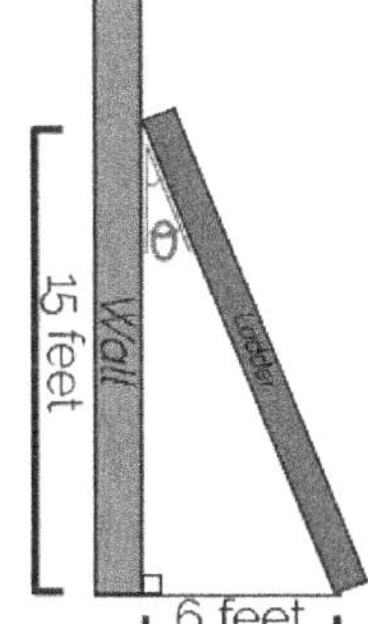

In this problem, we are looking to solve for a **missing angle**.

How can we solve this? We know we are trying to solve for (the unknown angle).

Let's redraw the problem.

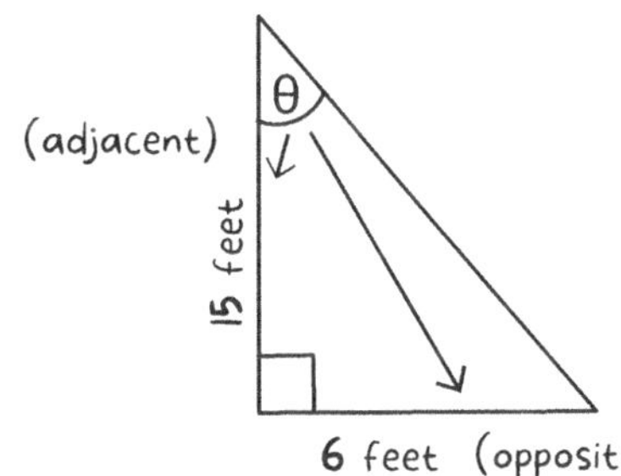

We know the **opposite** value is 6 and the **adjacent** value is 15 feet.

Which one of our trig functions deal with opposite and adjacent?

We need to use $\tan \theta = \frac{\text{opposite}}{\text{adjacent}}$. Plug in the known values.

$$\tan \theta = \frac{6}{15}$$

In this step, we must follow an important rule when using the calculator.

When **SOLVING** for **angles,** we must use the **inverse function** on the calculator.

The **inverse sine function** is $\sin^{-1}$. This can also be called **arcsin.**

The **inverse cos function** is $\cos^{-1}$. This can also be called **arccos.**

The **inverse tan function** is $\tan^{-1}$. This can also be called **arctan.**

The inverse sin takes the **ratio** of the respective trig function and gives us an **angle.**

We have $\tan\theta = \dfrac{6}{15}$ ($6 \div 15 = 0.4$ so...)

$\tan\theta = 0.4$ (We need to take the inverse tan function or arctan)

$\theta = \tan^{-1}(0.4)$ (Plug that into your calculator)

$\theta = 21.8°$

arctan(0.4

21.801409486

sin	cos	tan	◉Deg ○Rad		7	8	9	+	Back
sin⁻¹	cos⁻¹	tan⁻¹	π	e	4	5	6	–	Ans
xʸ	x³	x²	eˣ	10ˣ	1	2	3	×	M+
ʸ√x	³√x	√x	ln	log	0	.	EXP	/	M-
(	)	1/x	%	n!	±	RND	AC	=	MR

The angle between the ladder and the wall is 21.8°.

Example # 2

The distance from a boat to a lighthouse is **100 feet** and the lighthouse is **140 feet tall**. What is the angle of depression from the top of the Lighthouse to the boat?

We are provided with a word problem. Let's go ahead and solve this by drawing a diagram to assist us.

DAY 5
WEEK 5

MATH

SOLVING FOR MISSING ANGLES USING TRIG FUNCTIONS

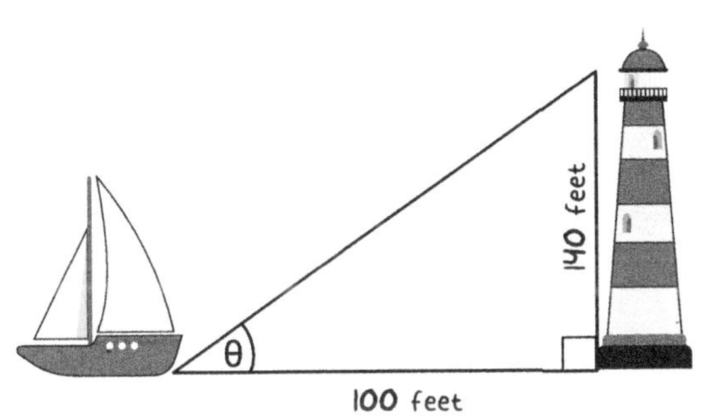

Your diagram should look like this. The angle of depression is the angle of elevation from the boat to the top of the light house.

How can we solve this? We know we are trying to solve for (the unknown angle).

We know the **opposite** value is 140 and the **adjacent** value is 100 feet.

Which one of our trig functions deal with opposite and adjacent?

We need to use $\tan \boldsymbol{\theta} = \frac{\text{opposite}}{\text{adjacent}}$. Plug in the known values.

$\tan \boldsymbol{\theta} = \frac{140}{100}$

$\tan \boldsymbol{\theta} = 1.4$ (We need to take the inverse tan function or arctan)

$\boldsymbol{\theta} = \tan^{-1}(1.4)$ (Plug that into your calculator)

$\boldsymbol{\theta} = 54.46°$ → Round that to the nearest tenth and we get $\boldsymbol{\theta} = 54.5°$. (Degrees are usually rounded to the nearest tenth unless the problem states otherwise).

Our answer is 54.5°.

Your turn to practice solving for missing angles using trig functions!

Directions: Find the measure of each **angle** indicated. Round your answer to the nearest tenth. Be sure to include the degree sign in your answer.

1.

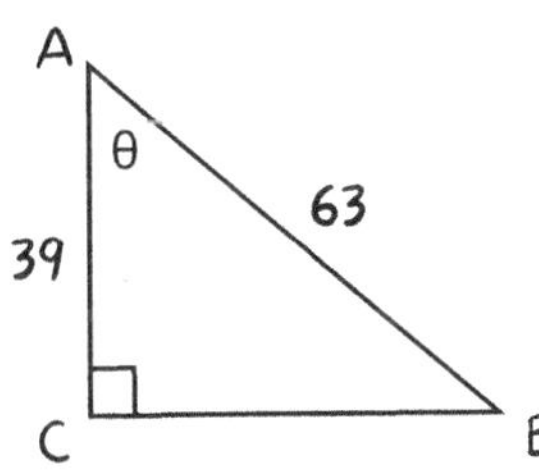

2.

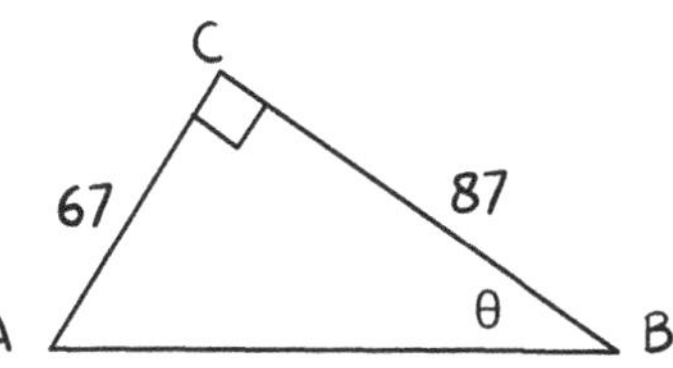

3.

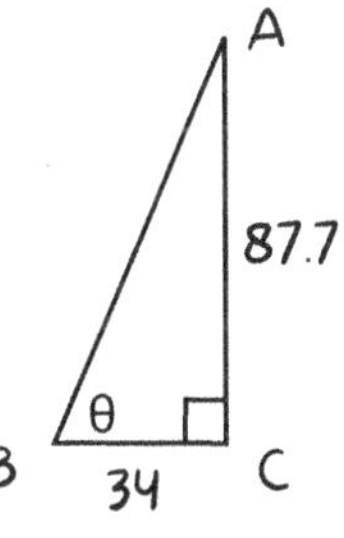

4.

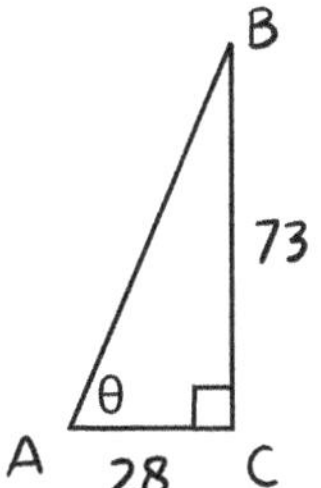

DAY 5

WEEK 5

MATH

SOLVING FOR MISSING ANGLES USING TRIG FUNCTIONS

5.

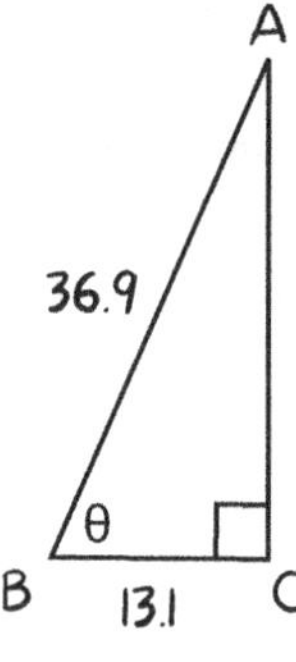

6.

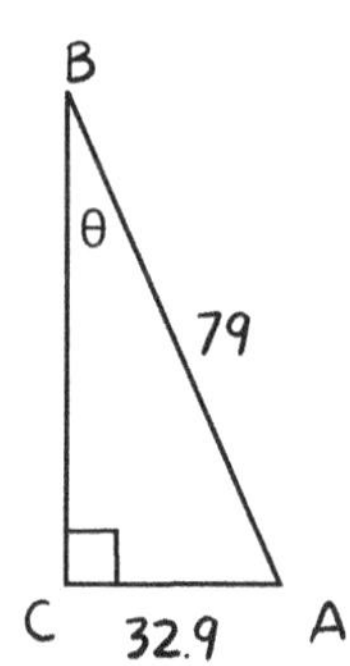

7.

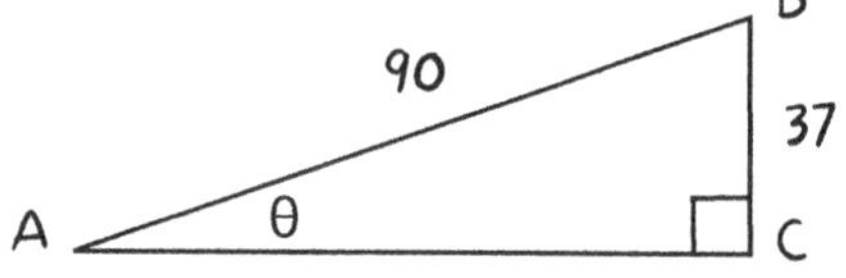

8.

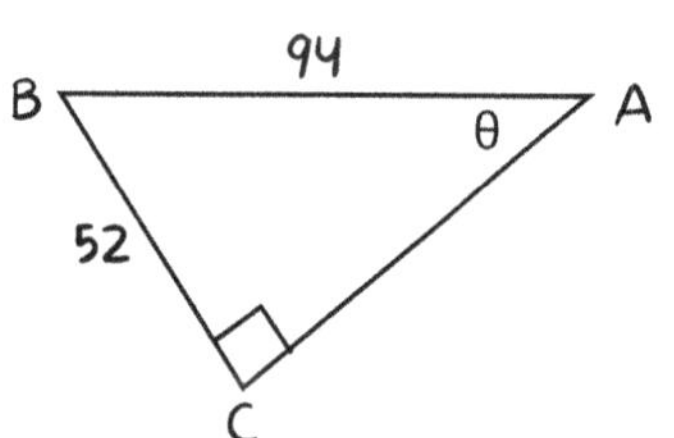

9.

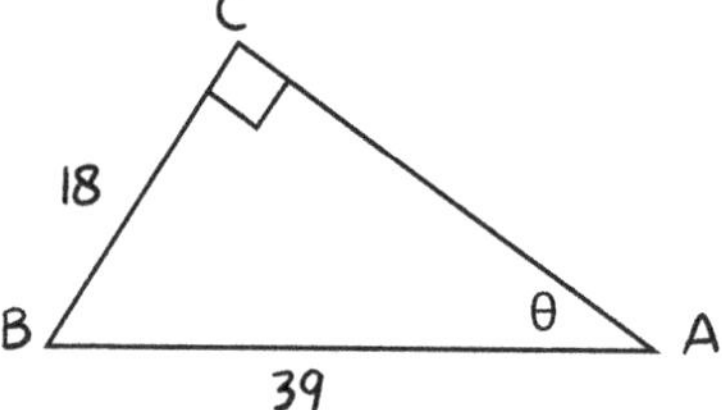

10.

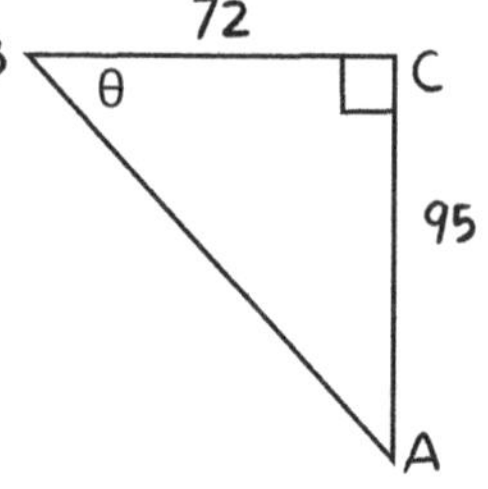

11.

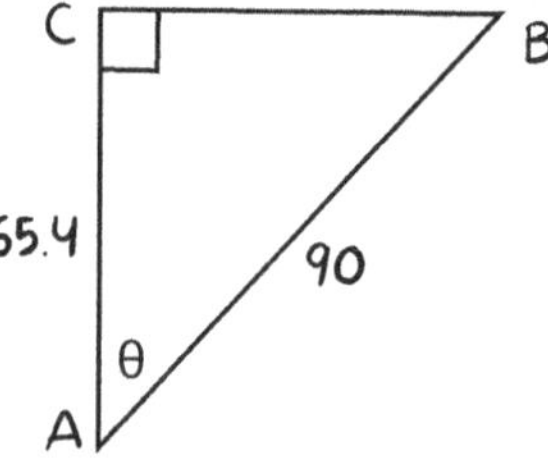

12.

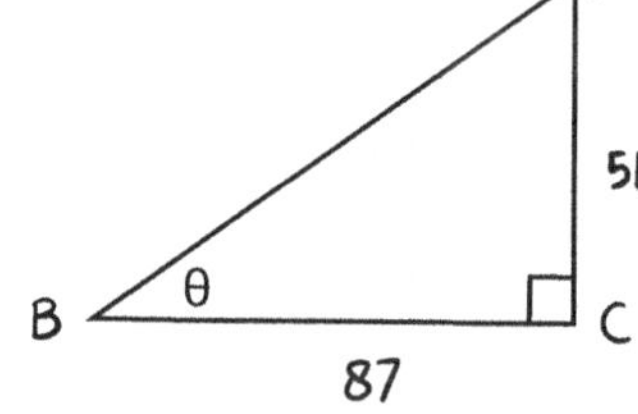

Let's get some fitness in! Go to page 169 to try some fitness activities.

DAY 6
WEEK 5

ELA
VOCABULARY

Directions: Read each sentence carefully and use context clues to determine the meaning of the underlined word. Choose the best definition for the underlined word from the options provided. Then, write a sentence using the word in a new context, demonstrating your understanding of its meaning and usage. You can use the Internet to look up the exact definition after answering the questions.

1. The politician's **mendacious** claims about his opponent's record were quickly exposed, causing a significant shift in public opinion.

A. honest

B. dull

C. false

D. persuasive

Sentence:

....

....

....

2. The **ephemeral** nature of the pop star's fame became evident when their latest album failed to generate much interest.

A. enduring

B. short-lived

C. undeserved

D. exaggerated

Sentence:

....

....

....

3. The persistent erosion of civil liberties has **inexorably** led to a state of widespread despondency amongst activists, which only galvanizes further protests.

A. gradually
B. noticeably
C. unpredictably
D. unavoidably

Sentence:

4. His propensity for **prolixity** often obscured the clarity of his arguments, causing many to disengage from what could potentially be insightful discussions.

A. brevity
B. verbosity
C. precision
D. curiosity

Sentence:

5. While her peers pursued lucrative careers in finance and technology, she was drawn to the **quixotic** challenge of reviving a failing newspaper.

A. unrealistic
B. practical
C. cautious
D. quick

Sentence:

Let's get some fitness in! Go to page 169 to try some fitness activities.

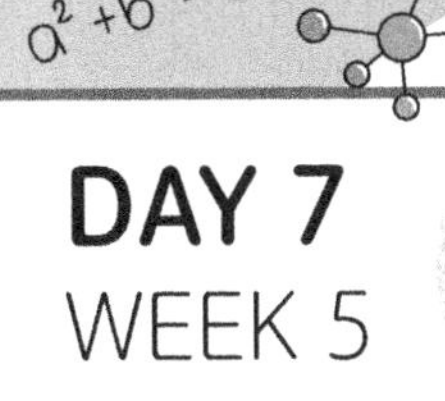

DAY 7
WEEK 5

SAT
ELA PREP

Directions: Work through these questions carefully.

1. The team at the Arctic Climate Research Lab is investigating the thickness of sea ice to evaluate whether recent temperature increases have ________ effect on ice stability, which could have serious implications for polar navigation and habitat preservation.

Which choice completes the text with the most logical and precise word or phrase?

A. a negligible

B. a profound

C. an unclear

D. a temporary

2. The gray wolf (Canis lupus) often displays complex social behavior, particularly when interacting within its pack, which includes showing submission by lowering ________ to the alpha.

Which choice completes the text so that it conforms to the conventions of Standard English?

A. them

B. themselves

C. it

D. itself

3. In the coral reefs off the coast of Belize, researchers have discovered that the coral species *Acropora palmata* and *Acropora cervicornis* contain a unique type of symbiotic algae known as *zooxanthellae*. These algae provide the corals with glucose, glycerol, and amino acids, which are crucial for their survival. In return, the corals provide the algae with protection and access to sunlight. The researchers hypothesize that this mutualistic relationship is vital for the corals' resilience to environmental stresses.

Which finding, if true, would most directly support the researchers' hypothesis?

A. *Acropora* corals can also survive in laboratory conditions without *zooxanthellae*.

B. *Acropora* corals are also found in deep sea areas where sunlight is limited.

C. Other coral species host different types of symbiotic organisms, not just algae.

D. Coral species without *zooxanthellae* have a lower survival rate under environmental stress conditions than *Acropora* species.

4. In a recent study on the migratory patterns of the monarch butterfly, biologists tracked the journey of these butterflies from North America to central Mexico. The researchers found that monarchs rely on the Earth's magnetic field to navigate during their migration. They hypothesize that the butterflies' antennae contain magnetic receptors that help them orient themselves relative to the Earth's magnetic field.

Which finding, if true, would most directly support the researchers' hypothesis?

A. Monarch butterflies exhibit similar navigation accuracy even when their antennae are covered with a non-magnetic material.

B. Monarch butterflies that were exposed to a distorted magnetic field in a controlled setting showed disorientation and difficulty in maintaining their migratory path.

C. The migratory paths of the monarch butterflies change slightly every year depending on the climate.

D. Other non-migratory butterfly species do not have the same structures in their antennae as monarch butterflies.

5. Recent observations of the blackcap bird (*Sylvia atricapilla*) in Europe show that some populations have started migrating northwest to the UK instead of southwest to Spain during the winter. Ornithologists hypothesize that this shift in migration patterns is due to warmer winter temperatures in the UK, making it a more hospitable environment during this season.

Which finding, if true, would most directly support the ornithologists' hypothesis?

A. Satellite tracking shows that blackcaps can fly longer distances without stopping for food or rest.

B. Blackcaps have been observed to change their diet preferences when in the UK.

C. Blackcaps that migrate to the UK have a higher survival rate and reproductive success compared to those migrating to Spain.

D. Genetic studies show that blackcaps in the UK and those in Spain are becoming increasingly genetically distinct.

Let's get some fitness in! Go to page 169 to try some fitness activities.

WEEK 6

GRADE 10-11

This week, you will read passages that are accompanied by a data table. This type of skill (analyzing data in passages) is regularly tested on standardized exams. For math, you will learn about the 30-60-90 degree and 45-45-90 degree rules.

ELA

ANALYZING DATA IN ELA PASSAGES

Look at the chart below and read the passage. Then answer the question.

Country	Gini Coefficient	Human Development Index (HDI)	Education Index	Life Expectancy at Birth (years)
Norway	0.27	0.957	0.919	82.4
USA	0.41	0.926	0.900	78.9
Brazil	0.53	0.765	0.686	75.7
China	0.38	0.761	0.656	76.7
South Africa	0.63	0.709	0.724	64.1

The data in the chart above presents a comparison of four key socioeconomic indicators across five countries: Norway, the United States, Brazil, China, and South Africa. These indicators shed light on the varying levels of inequality, human development, education, and health outcomes in these nations.

The Gini coefficient, which measures income inequality within a country, ranges from 0 (perfect equality) to 1 (perfect inequality). Among the countries listed, South Africa has the highest Gini coefficient at 0.63, indicating a highly unequal distribution of income. This can be attributed to factors such as the legacy of apartheid, high unemployment rates, and limited access to education and economic opportunities for a significant portion of the population. In contrast, Norway has the lowest Gini coefficient at 0.27, reflecting a more equal distribution of income, likely due to progressive taxation, a strong social welfare system, and policies aimed at reducing poverty and promoting social mobility.

The Human Development Index (HDI) is a composite measure of a country's average achievements in three key dimensions: health, education, and standard of living. Norway ranks highest among the countries listed, with an HDI of 0.957, indicating a very high level of human development. This can be attributed to factors such as a high-quality education system, universal healthcare, and a high standard of living. In contrast, South Africa has the lowest HDI at 0.709, suggesting lower levels of human development, likely due to challenges such as poverty, inequality, and limited access to quality education and healthcare.

DAY 1
WEEK 6

ELA
ANALYZING DATA IN ELA PASSAGES

The Education Index, a component of the HDI, measures a country's average educational attainment and expected years of schooling. Norway and the United States have the highest Education Index values at 0.919 and 0.900, respectively, reflecting strong education systems and high levels of educational attainment. Brazil and China have lower Education Index values at 0.686 and 0.656, respectively, indicating potential challenges in access to quality education and lower levels of educational attainment.

Life expectancy at birth is a key indicator of a country's overall health and well-being. Norway has the highest life expectancy at 82.4 years, reflecting factors such as high-quality healthcare, healthy lifestyles, and low levels of poverty. South Africa has the lowest life expectancy at 64.1 years, likely due to challenges such as high rates of infectious diseases, limited access to healthcare, and socioeconomic disparities.

Question:

Based on the information provided in the data chart and passage, which of the following statements best describes the relationship between the Gini coefficient and the Human Development Index (HDI) for the countries listed?

A. Countries with a lower Gini coefficient consistently have a higher HDI, indicating that greater income equality is strongly associated with higher levels of human development.

B. There is a weak positive correlation between the Gini coefficient and HDI, suggesting that income inequality has a limited impact on human development.

C. The relationship between the Gini coefficient and HDI is inconsistent, with some countries demonstrating high levels of human development despite high income inequality.

D. The data provided is insufficient to draw a definitive conclusion about the relationship between the Gini coefficient and HDI across the countries listed.

Let's get some fitness in! Go to page 169 to try some fitness activities.

DAY 2
WEEK 6

ELA
ANALYZING DATA IN ELA PASSAGES

Directions: Read the passage below. Then answer the questions.

Country	GDP per capita (USD)	Life Expectancy (years)	Human Development Index
A	45,000	82.3	0.920
B	62,000	81.7	0.935
C	39,000	83.1	0.915
D	55,000	80.9	0.925
E	51,000	82.8	0.930

The data chart compares five countries across three different metrics: GDP per capita, life expectancy, and Human Development Index (HDI). GDP per capita is a measure of a country's economic output per person, while life expectancy reflects the average number of years a person is expected to live. The HDI is a composite statistic that takes into account a country's GDP per capita, life expectancy, and education levels to provide a more comprehensive measure of a country's development.

Among the five countries, Country B has the highest GDP per capita at $62,000, followed by Country D at $55,000. Country C has the lowest GDP per capita at $39,000. However, when considering life expectancy, Country C ranks the highest at 83.1 years, while Country D has the lowest life expectancy at 80.9 years. In terms of HDI, Country B has the highest score at 0.935, while Country C has the lowest at 0.915.

Question:

Which of the following statements is best supported by the information provided in the data chart and passage?

A. A higher GDP per capita always correlates with a higher life expectancy.

B. The country with the highest HDI score also has the highest life expectancy.

C. GDP per capita is the most important factor in determining a country's HDI score.

D. A country's development cannot be fully assessed by considering only one metric.

DAY 3
WEEK 6

MATH
SOLVING FOR MISSING ANGLES USING TRIG FUNCTIONS

Directions: Find the measure of each **angle** indicated. Round your answer to the nearest tenth. Be sure to include the degree sign in your answer.

1.

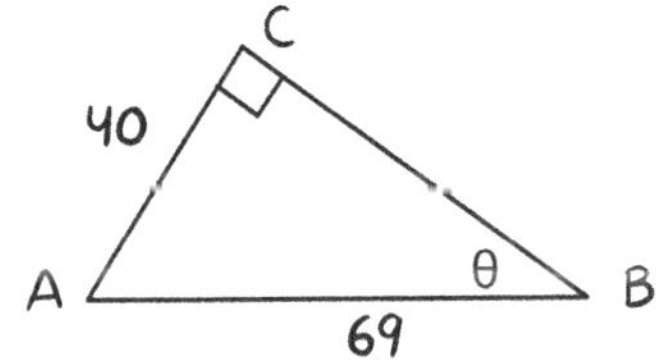

2.

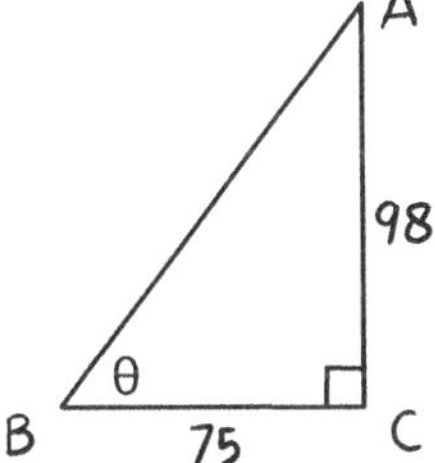

3.

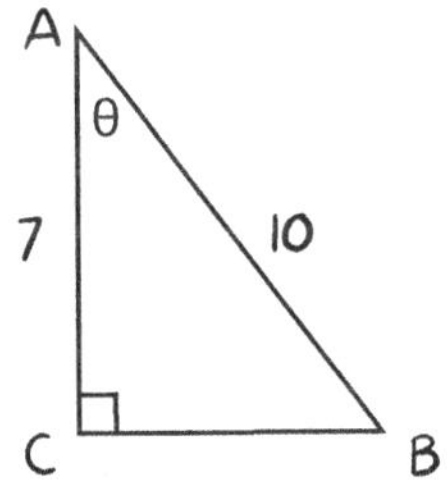

4.

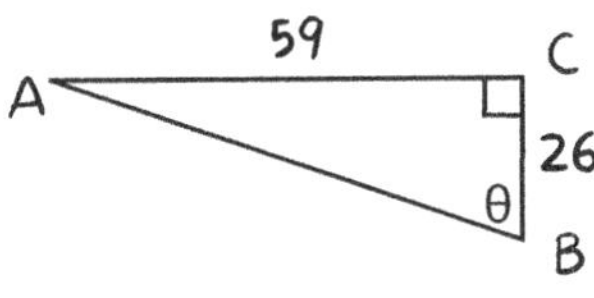

5.

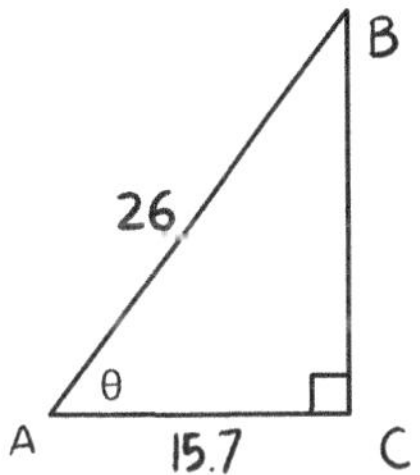

6.

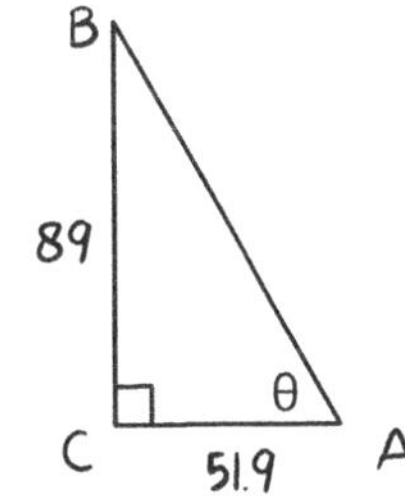

7.

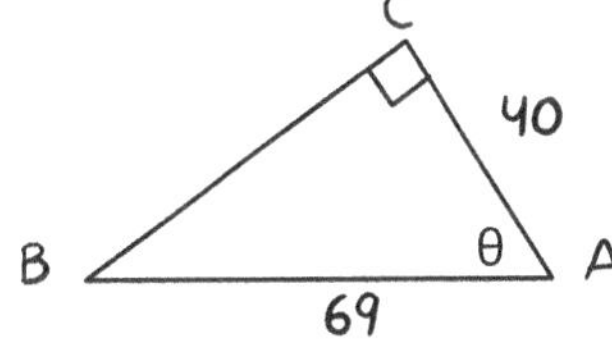

8.

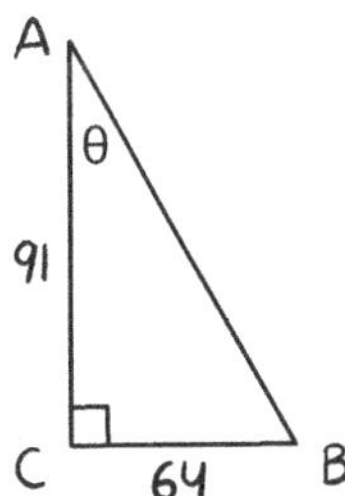

DAY 3
WEEK 6

MATH

SOLVING FOR MISSING ANGLES USING TRIG FUNCTIONS

9.

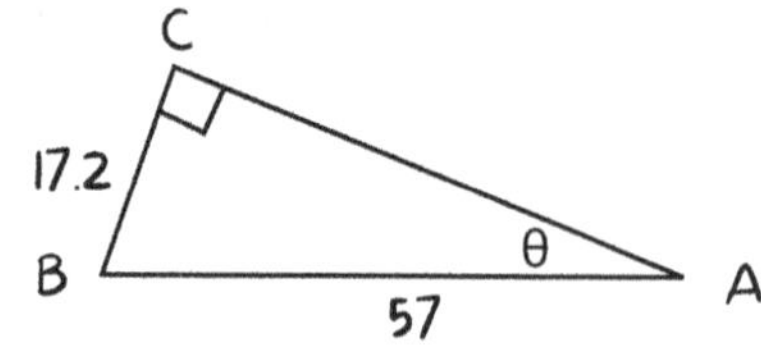

10.

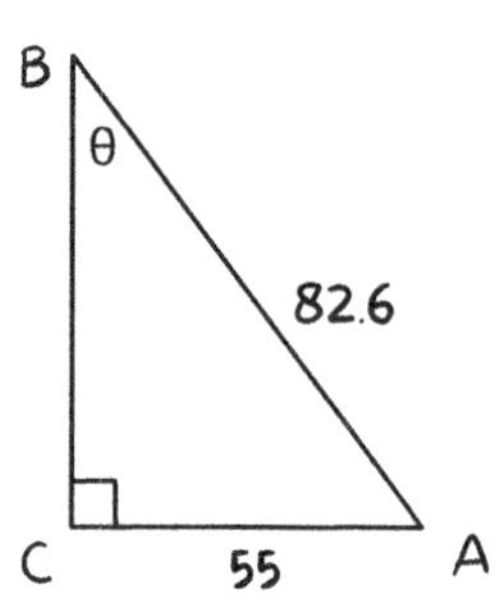

11.

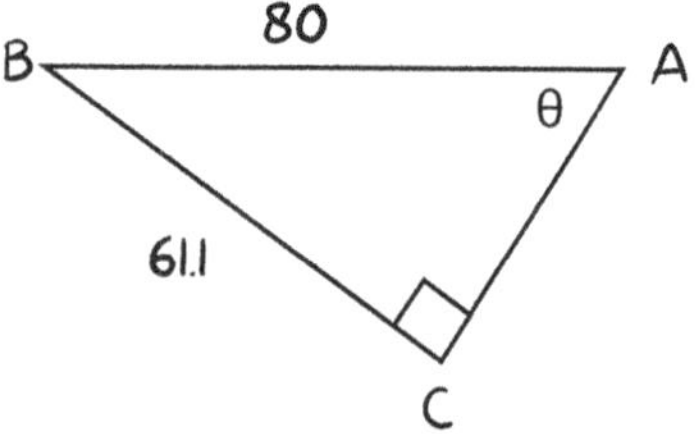

12.

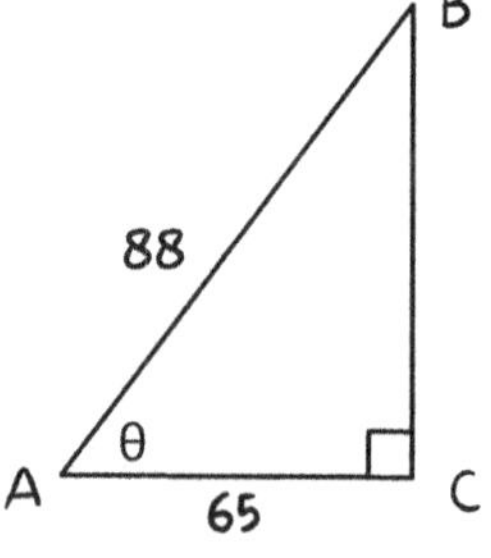

13.

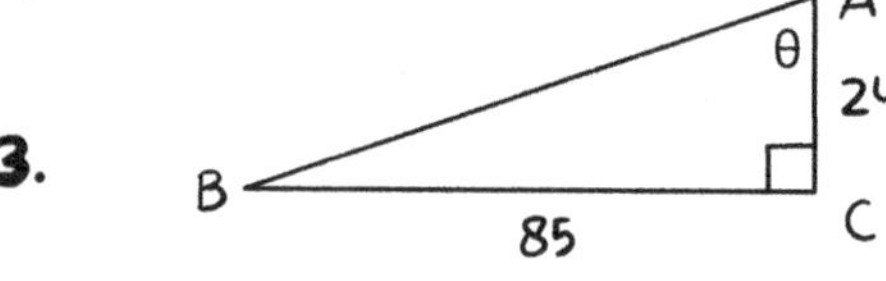

14.

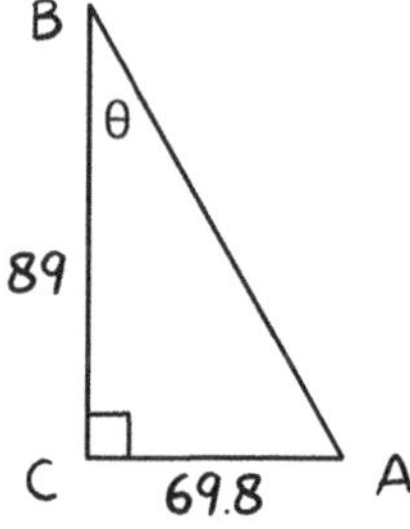

15.

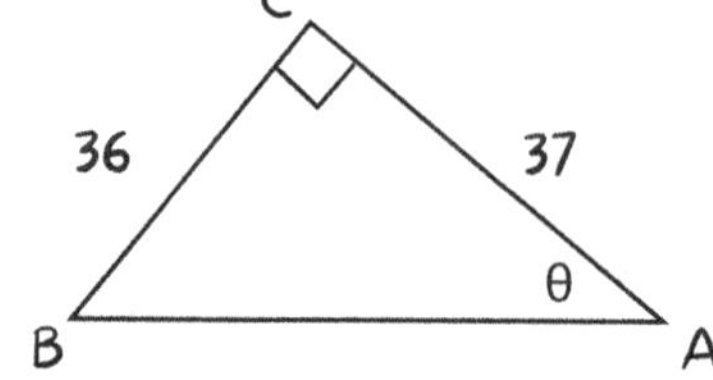

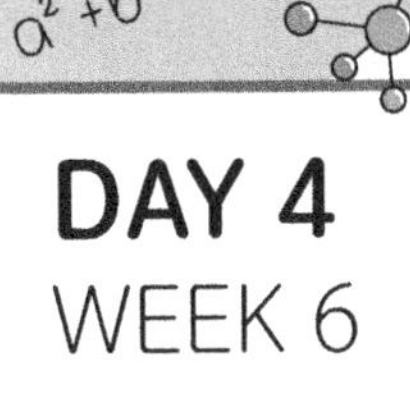

DAY 4
WEEK 6

MATH
SPECIAL ANGLES

OVERVIEW:

The first special right triangle we will cover is the **30-60-90** triangle that has the angles 30°, 60°, and 90°. A 30-60-90 right triangle has the side ratios x, $x\sqrt{3}$, $2x$.

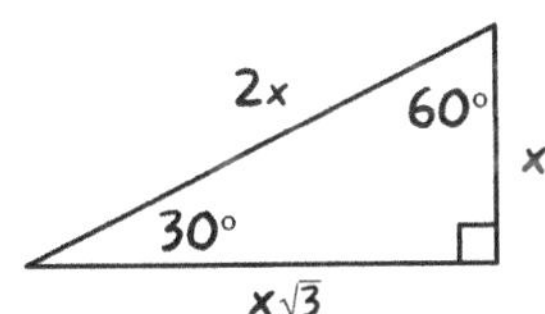

The side opposite of 30° is always the shortest side (x).

The side opposite of 60° is the longer leg ($x\sqrt{3}$).

The hypotenuse is always $2x$.

Knowing this ratio, we can find the value of 2 missing sides if we are provided with just **one** side value. Let's take a look at a few examples.

Example # 1

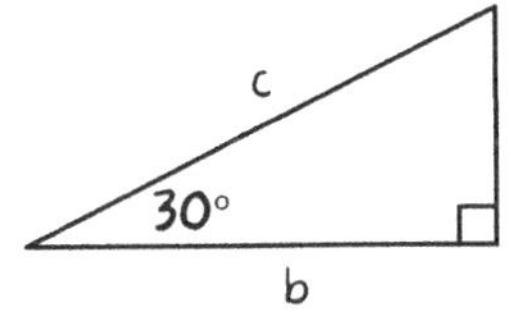

This is a 30-60-90 triangle. We know the value of the shorter leg (x) is 5.

b is $x\sqrt{3}$, but we know x is 5 so b is $5\sqrt{3}$.

c is the hypotenuse and is always $2x$. We know x is 5, so $2x$ is 10.

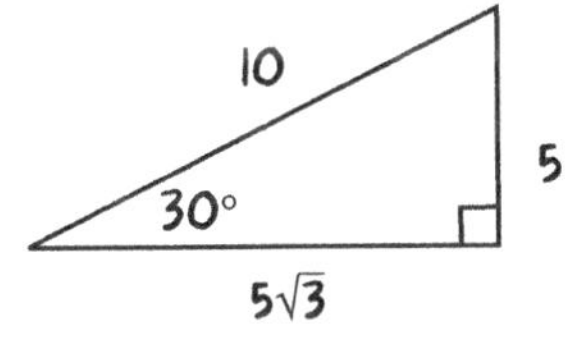

b = $5\sqrt{3}$ and c = 10

Example 2:

Find the length of the hypotenuse in the figure below.

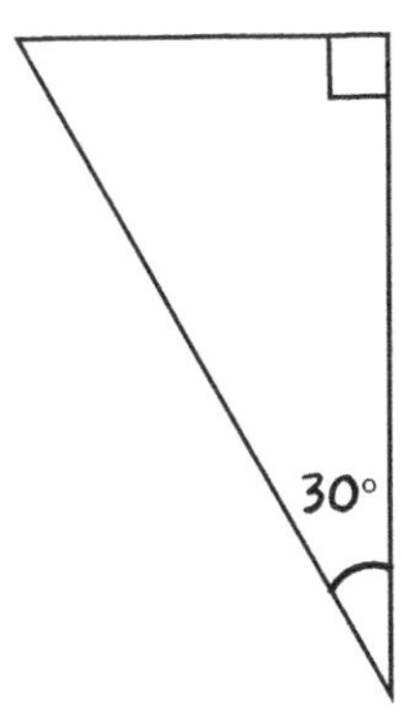

This is a 30-60-90 triangle. Opposite of 30°, we know the ratio is x. Opposite of 60° is $x\sqrt{3}$.

We can see that 18 is equal to $x\sqrt{3}$. If we want to figure out what x is, what can we do? We can divide by $\sqrt{3}$.

So $x = \frac{18}{\sqrt{3}}$, which is the shortest leg measure.

However, we must remember that we cannot have a square root in our denominator. We must rationalize this denominator.

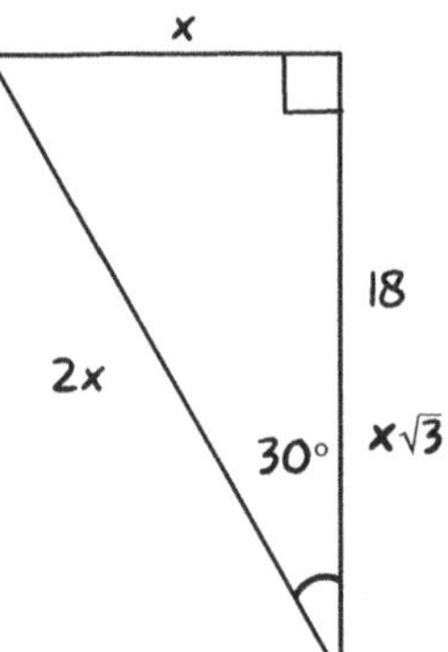

$$\frac{18}{\sqrt{3}} \times \frac{\sqrt{3}}{\sqrt{3}} = \frac{18\sqrt{3}}{3} = 6\sqrt{3}$$

So $x = 6\sqrt{3}$

We can easily find the hypotenuse now which is $2x$. $2(6\sqrt{3}) = 12\sqrt{3}$. The hypotenuse of the figure above is $12\sqrt{3}$.

The second special right triangle we will cover is the **45-45-90** triangle that has the angles 45°, 45°, and 90°. A 45-45-90 right triangle has the side ratios x, x, $x\sqrt{2}$.

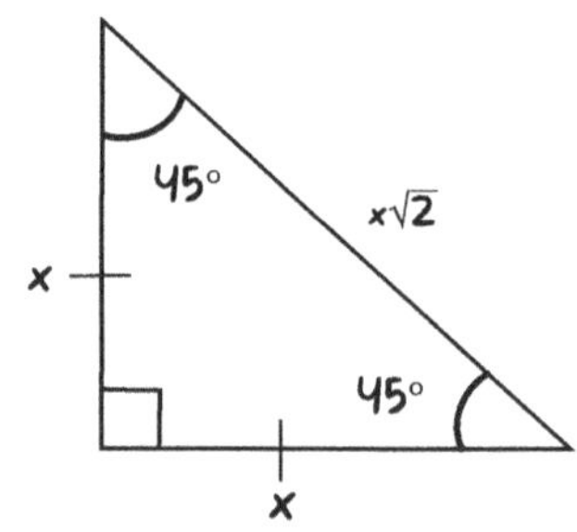

The hypotenuse always has the ratio of $x\sqrt{2}$ and the legs are x. Let's take a look at a few examples.

Example 1:

What is the value of u and v?

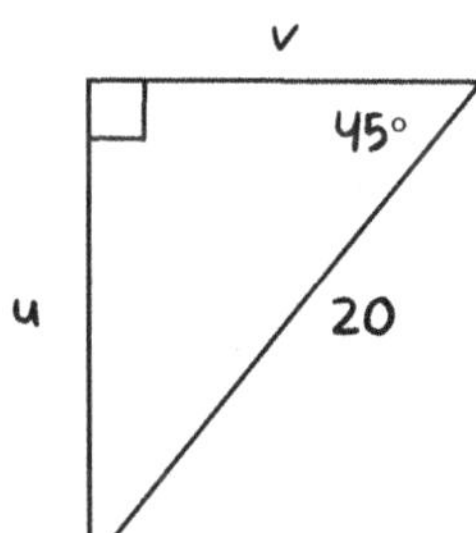

In a 45-45-90 right triangle, the legs are congruent, and the hypotenuse is $\sqrt{2}$ times longer than one of the legs.

Since the hypotenuse is 20, in order to solve for the length of a leg, we will divide by $\sqrt{2}$.

$\frac{20}{\sqrt{2}} = \frac{20}{\sqrt{2}} \times \frac{\sqrt{2}}{\sqrt{2}} = \frac{20\sqrt{2}}{2} = 10\sqrt{2}$ ← We rationalized the denominator and simplified.

The value of $u = 10\sqrt{2}$.

The value of $v = 10\sqrt{2}$.

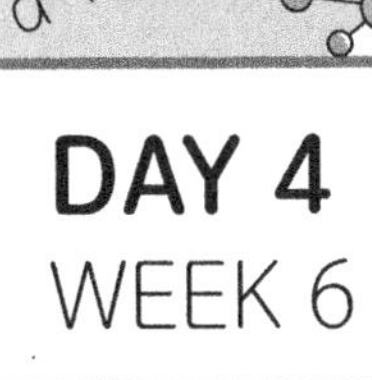

DAY 4
WEEK 6

MATH
SPECIAL ANGLES

Example 2:

What is the value of x and y?

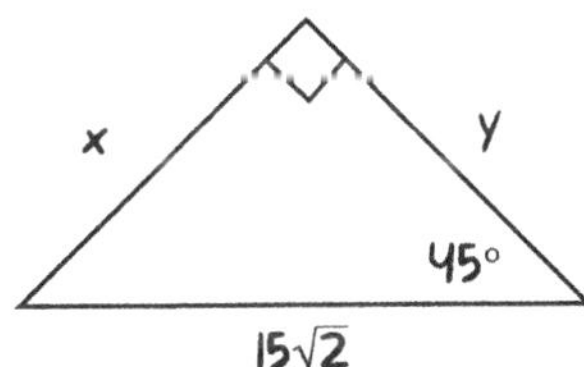

In a **45 45 90** right triangle, the legs are congruent, and the hypotenuse is $\sqrt{2}$ times longer than one of the legs.

Since the hypotenuse is $15\sqrt{2}$, in order to solve for the length of a leg, we will divide by $\sqrt{2}$.

$\frac{15\sqrt{2}}{\sqrt{2}} = 15$. So the value of each leg is 15.

The value of $x = 15$.

The value of $y = 15$.

Your turn to practice special angle problems!

Directions: The following problems are a **30-60-90** right triangle or a **45-45-90** right triangle. Solve for the variables asked. Do **not** use a calculator. Make sure your answers are in the simplest form.

1.

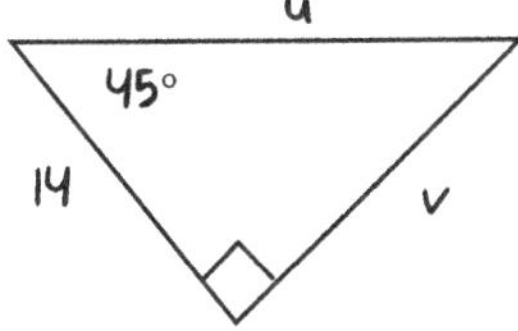

2.

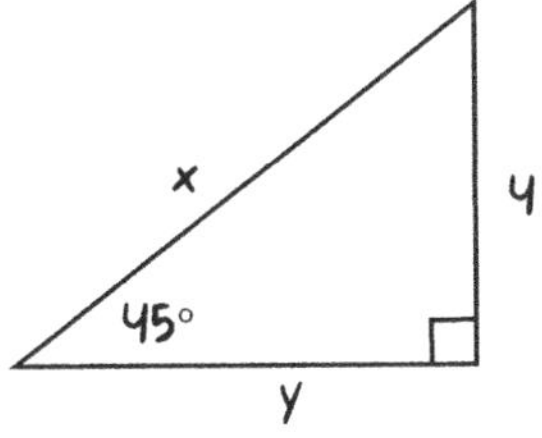

3.

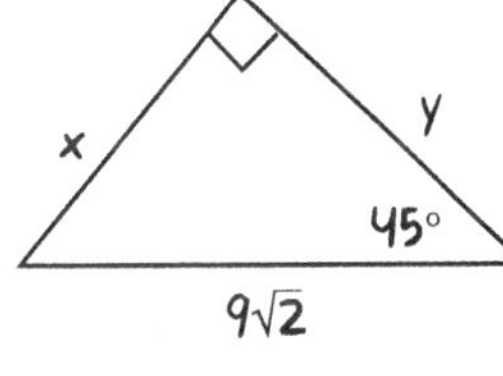

4.

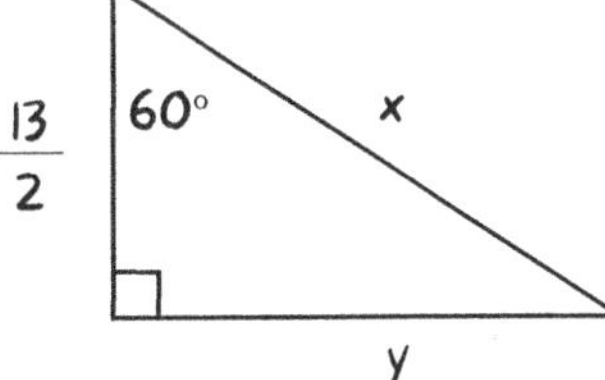

MATH
SPECIAL ANGLES

5.

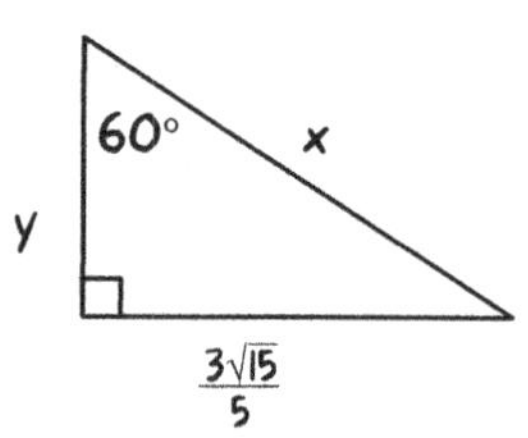

8.

u
v
60°
$\frac{7\sqrt{2}}{2}$

6.

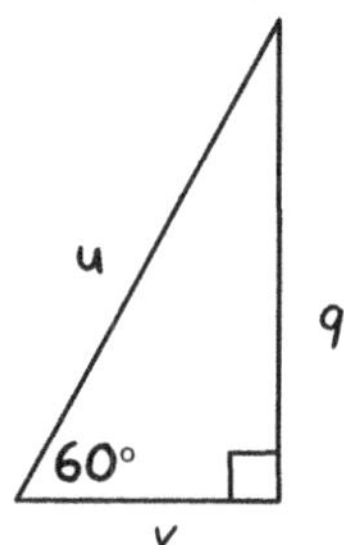

9.

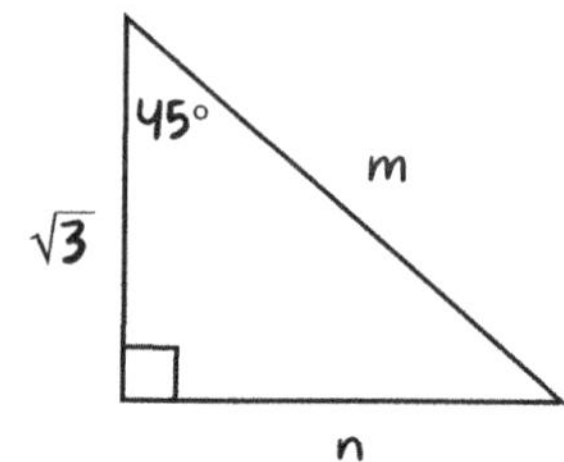

7.

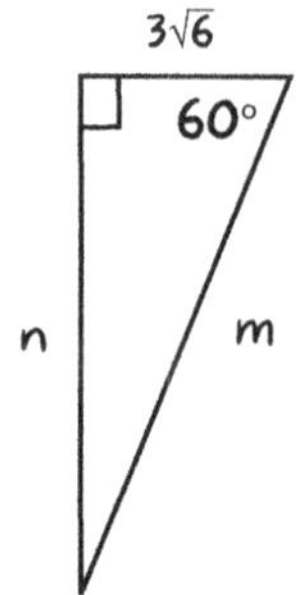

10.

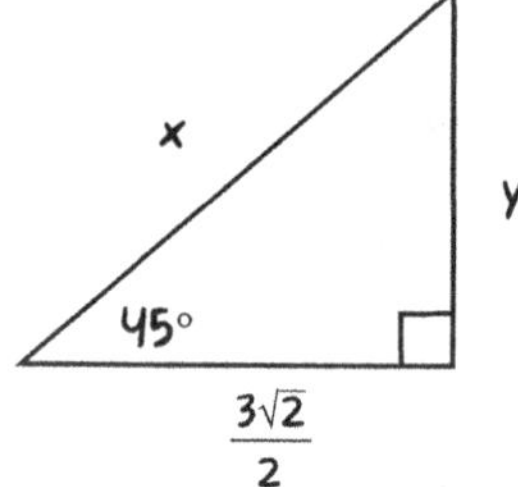

Let's get some fitness in! Go to page **169** to try some fitness activities.

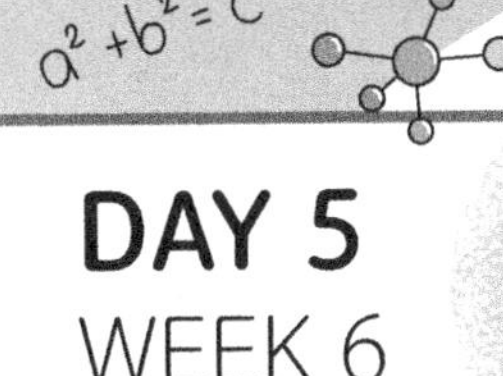

DAY 5
WEEK 6

MATH
SPECIAL ANGLES

On Day 3, you solved problems dealing with the two special right triangles and their ratios. Let's take a look at these two examples below for today.

Example 1:

What is the value of x in the diagram below?

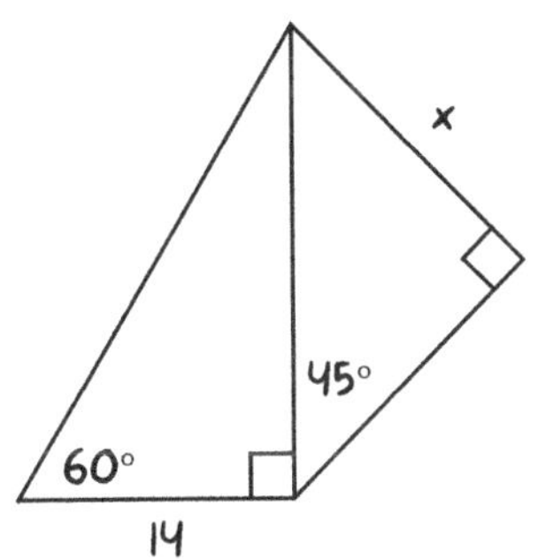

We need to figure out the value of x. We can see in this diagram we have **two** right triangles, one of them being the 30-60-90 triangle and the other a 45-45-90 triangle.

In order to figure out the value of x, we need to figure out the value of the hypotenuse on the 45-45-90 triangle.

Notice the hypotenuse of the 45-45-90 triangle is the **leg** for the 30-60-90 triangle.

Let's recall our ratios for a 30-60-90 triangle.

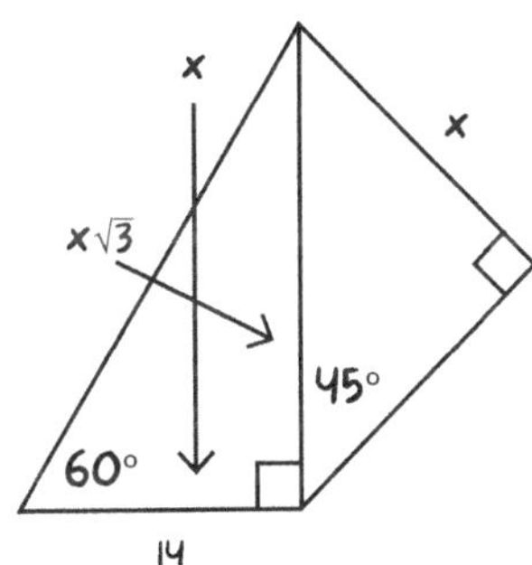

14 is the value of "x" in the 30-60-90 right triangle.

We can easily determine the other leg value, as we know the ratio is $x\sqrt{3}$. The value is $14\sqrt{3}$.

Now let's re-draw just the 45-45-90 triangle.

Now we can use our ratios that we know for this 45-45-90 triangle.

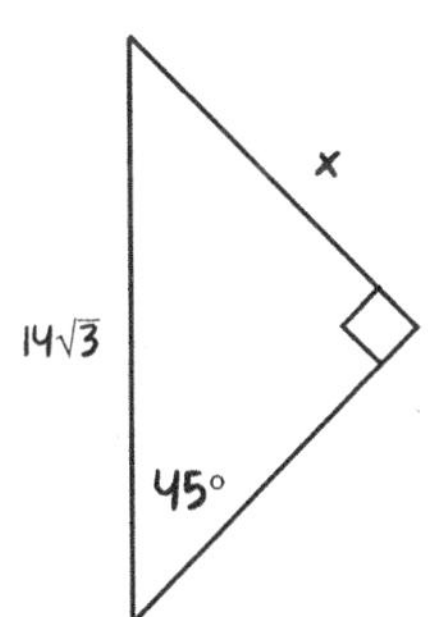

We know a 45-45-90 right triangle has the side ratios x, x, $x\sqrt{2}$. $14\sqrt{3}$ represents the ratio $x\sqrt{2}$. How can we identify what x is? Simply divide by $\sqrt{2}$.

$$\frac{14\sqrt{3}}{\sqrt{2}} = \frac{14\sqrt{3}}{\sqrt{2}} \times \frac{\sqrt{2}}{\sqrt{2}} = \frac{14\sqrt{6}}{2} = 7\sqrt{6}$$

Our answer to this problem is $7\sqrt{6}$.

Example 2:

What is the value of y in the diagram below?

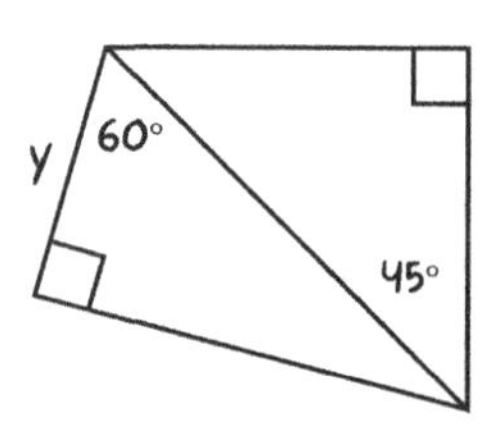

We need to figure out the value of x. We can see in this diagram we have **two** right triangles, one of them being the 45-45-90 triangle and the other a 30-60-90 triangle.

We need to first tackle the 45-45-90 right triangle because we were provided with the leg value. We know a 45-45-90 right triangle has the side ratios x, x, $x\sqrt{2}$.

We know the leg value $8\sqrt{6}$ represents the side ratio x. We can determine the hypotenuse by multiplying $8\sqrt{6}$ by $\sqrt{2}$.

$8\sqrt{6} \times \sqrt{2} = 8\sqrt{12} = 8\sqrt{4}\sqrt{3} = (8)(2)\sqrt{3} = 16\sqrt{3}$

The hypotenuse is $16\sqrt{3}$.

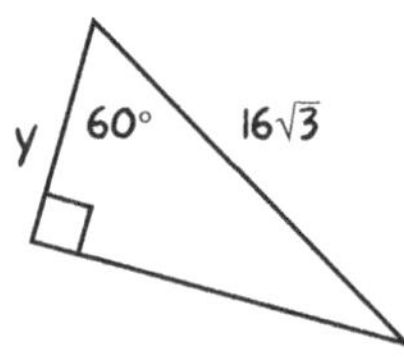

Let's re-draw the 30-60-90 right triangle with the new information we know.

A 30-60-90 right triangle has the side ratios x, $x\sqrt{3}$, $2x$.

Let's also label this with our known ratios.

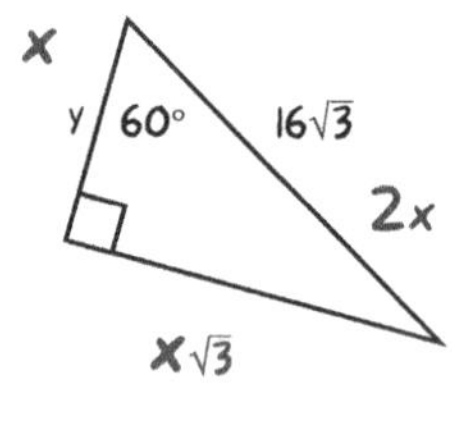

We need to figure out what the variable y is. We can see y represents the ratio x, and we know the hypotenuse, $16\sqrt{3}$, which represents the ratio 2x.

What do we need to do? Simply divide $16\sqrt{3}$ by 2.

$$\frac{16\sqrt{3}}{2} = 8\sqrt{3}$$

Our answer to this problem is $8\sqrt{3}$.

Your turn to practice special angle problems!

DAY 5
WEEK 6

MATH
SPECIAL ANGLES

Directions: Find the missing side length. Do **not** use a calculator. Make sure your answers are in the simplest form.

1.

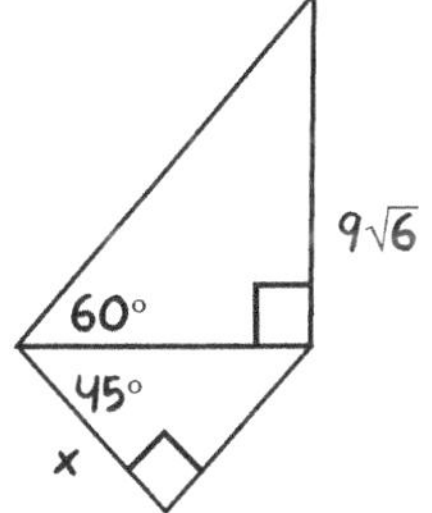

2.

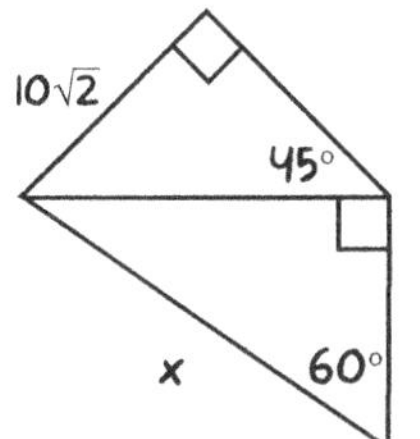

3.

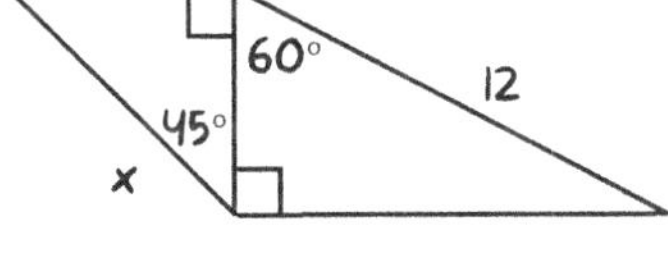

4.

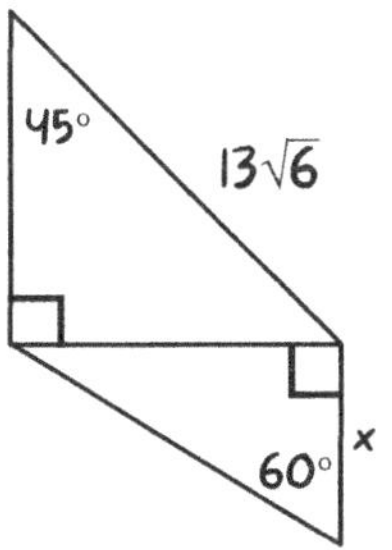

5.

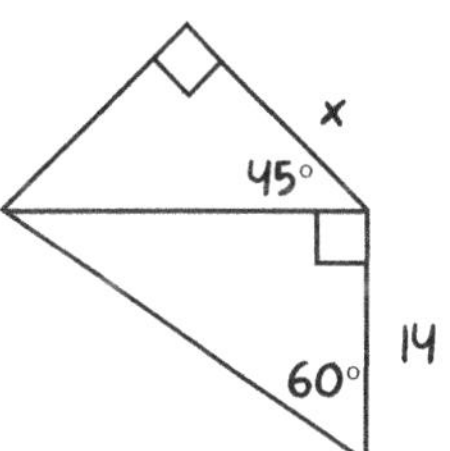

6.

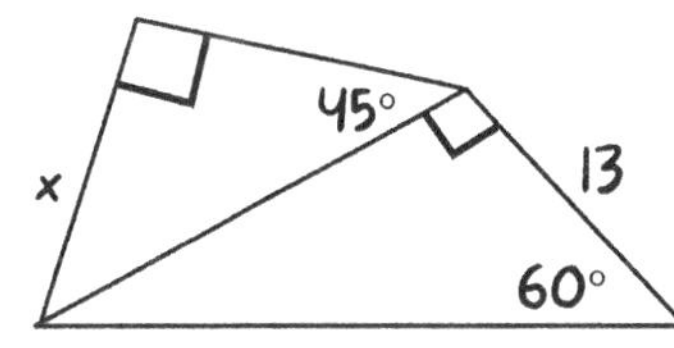

7.

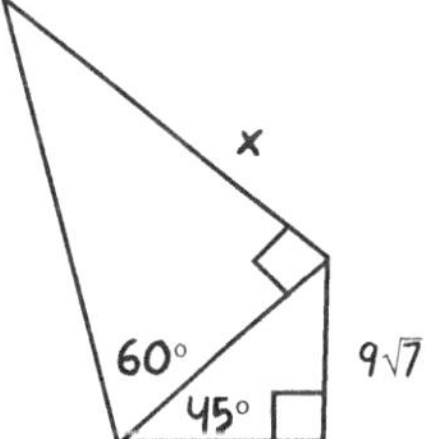

8.

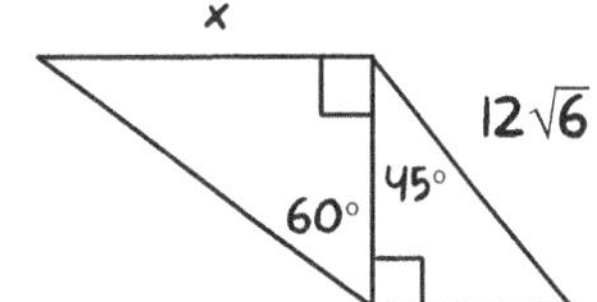

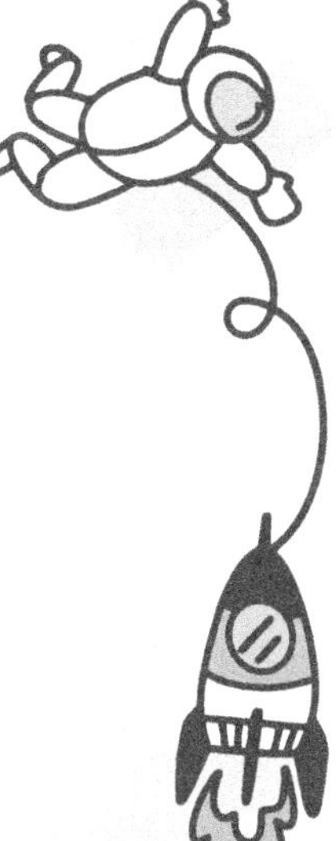

MATH

SPECIAL ANGLES

9.

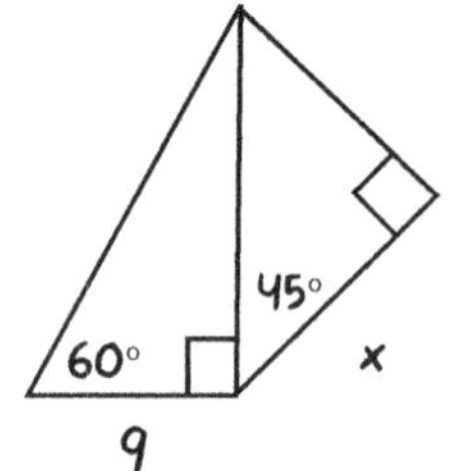

10.

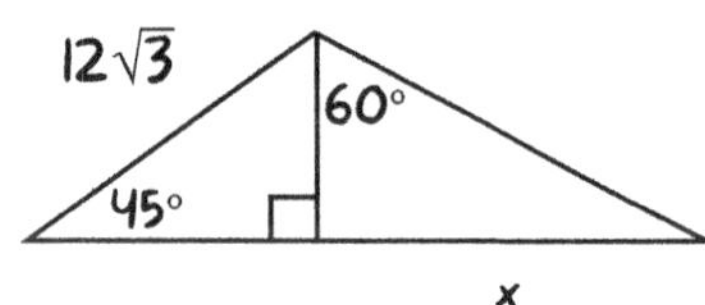

11.

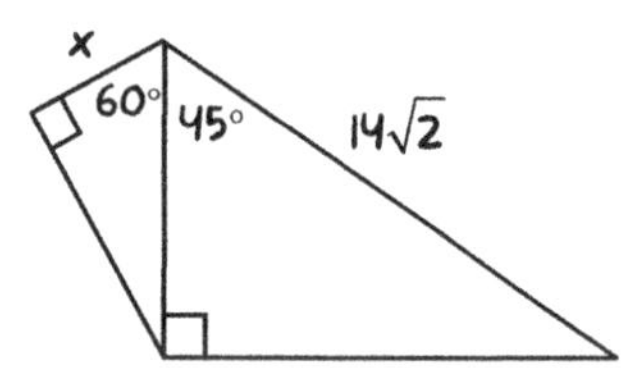

12.

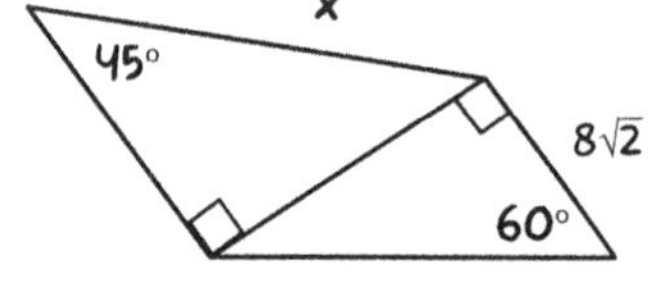

13.

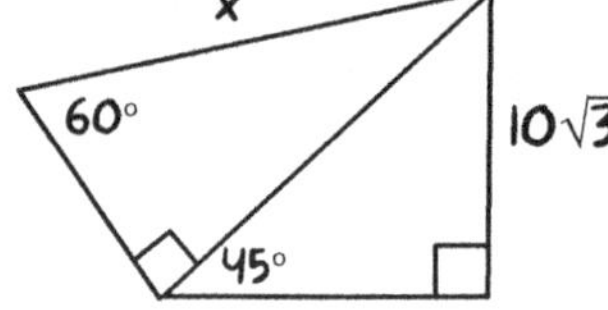

14.

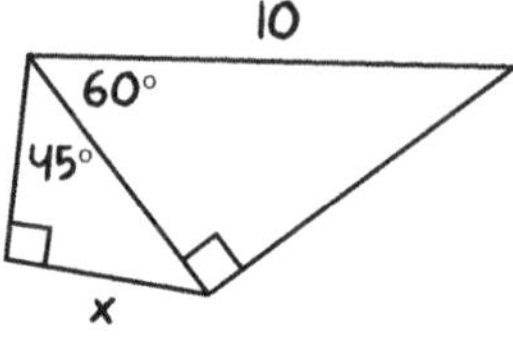

15.

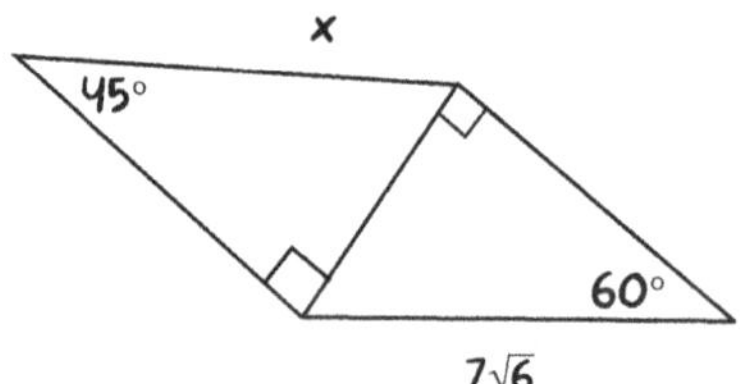

Let's get some fitness in! Go to page 169 to try some fitness activities.

ELA
VOCABULARY

Directions: Read each sentence carefully and use context clues to determine the meaning of the underlined word. Choose the best definition for the underlined word from the options provided. Then, write a sentence using the word in a new context, demonstrating your understanding of its meaning and usage. You can use the Internet to look up the exact definition after answering the questions.

1. In her latest series of essays, the author delves into the **arcane** rituals of ancient civilizations, shedding light on practices that remain largely misunderstood by modern scholars.

A. well-known
B. mysterious
C. simple
D. mainstream

Sentence:

....

....

....

2. His decision to incorporate **nefarious** tactics in the corporate strategy not only tarnished his reputation but also led to significant legal repercussions.

A. beneficial
B. unimportant
C. wicked
D. direct

Sentence:

....

....

....

3. His mastery in crafting **surreptitious** plans ensured that the team often had the upper hand in negotiations without the opposition realizing it.

A. overt
B. clumsy
C. secretive
D. straightforward

Sentence:

....

....

....

4. The documentary highlighted the **transient** nature of fame and how quickly public opinion can change in the entertainment industry.

A. lasting
B. infrequent
C. constant
D. fleeting

Sentence:

....

....

....

5. The old manor's **derelict** state made it an eerie yet fascinating subject for photographers and urban explorers.

A. abandoned
B. crowded
C. renovated
D. vibrant

Sentence:

....

....

....

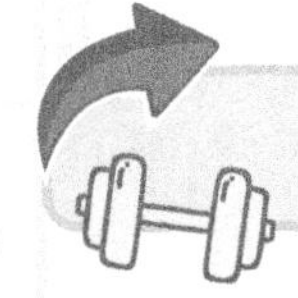

Let's get some fitness in! Go to page 169 to try some fitness activities.

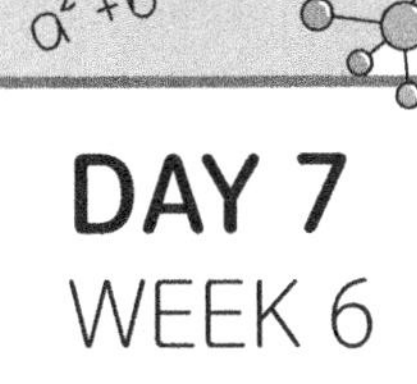

DAY 7 WEEK 6

SAT ELA PREP

Directions: Work through these questions carefully.

1. The Hubble Space Telescope, launched into low Earth orbit in 1990, has been instrumental in many astronomical discoveries due to its ability to capture high-resolution images of distant galaxies, nebulae, and other celestial objects. Its success has ___________ the development of even more advanced space telescopes.

Which choice completes the text with the most logical and precise word or phrase?

A. hindered

B. inspired

C. delayed

D. overshadowed

2. The novel "To Kill a Mockingbird" by Harper Lee has been widely celebrated for its powerful themes of racial injustice and the loss of innocence. The story, narrated by a young girl named Scout, ___________ the reader to view the world through the lens of a child confronting the harsh realities of prejudice and inequality.

Which choice completes the text so that it conforms to the conventions of Standard English?

A. compel

B. compelled

C. compelling

D. compels

3. In a groundbreaking study, archaeologists have discovered evidence of advanced agricultural practices in the ancient Maya civilization. The researchers found that the Maya utilized a sophisticated system of canals, reservoirs, and raised fields to maximize crop yields in their tropical environment. This finding challenges the previous notion that the Maya relied solely on slash-and-burn agriculture.

Which finding, if true, would most directly support the researchers' claim?

A. The Maya civilization collapsed due to prolonged droughts and environmental degradation.

B. Similar agricultural systems have been found in other ancient civilizations in Central and South America.

C. The Maya also developed advanced astronomical knowledge and a complex calendar system.

D. Pollen analysis from ancient Maya fields shows a diverse array of crops, including maize, squash, and beans.

4. Psychologists studying the concept of "flow" have found that individuals experiencing this state exhibit heightened focus, creativity, and enjoyment while engaged in a challenging task. They hypothesize that flow occurs when a person's skills are evenly matched with the difficulty of the task at hand.

   Which scenario would most directly illustrate the psychologists' hypothesis?

   A. An experienced rock climber becomes fully immersed in the challenge of ascending a difficult route.

   B. A skilled musician struggles to learn a new, highly complex piece of music.

   C. A professional chess player easily defeats a novice opponent in a tournament.

   D. A talented artist becomes bored while completing a simple, repetitive drawing exercise.

5. The concept of "umami," a savory taste distinct from sweet, salty, sour, and bitter, has been gaining recognition in the culinary world. Umami is often associated with the presence of glutamates, which are naturally occurring in many foods, such as ripe tomatoes, aged cheeses, and fermented products. Some scientists argue that the human taste receptors for umami evolved as a way to detect the presence of proteins and amino acids in food, ____________ the nutritional value of a particular food item.

   Which choice completes the text so that it conforms to the conventions of Standard English?

   A. a signal

   B. signaling

   C. to signal

   D. having signaled

Let's get some fitness in! Go to page **169** to try some fitness activities.

WEEK 7

GRADE 10-11

This week, you will read passages and answer the questions for the ELA section. For math, you will learn about Law of Sines, Law of Cosines, and the arithmetic sequence. As usual, the last day of each week is reserved for SAT practice on the ELA section.

DAY 1
WEEK 7

ELA
READING COMPREHENSION PASSAGE

Directions: Read the passage below. Then answer the questions.

A Walk in the Woods

Henry David Thoreau, the 19th century American essayist and philosopher, once wrote "I went to the woods because I wished to live deliberately." This impulse to seek wisdom and peace in nature has been shared by many thoughtful individuals over the centuries. As our world grows evermore chaotic and distracting, the simple act of taking a walk in the woods provides a much-needed respite and opportunity for introspection.

Whenever I find myself growing weary of the noise and haste of modern life, I make my way to a quiet wooded trail not far from my home. As soon as I step out of my car and set foot on the soft earth path, I feel my spirit begin to lift. The farther I walk into the welcoming arms of the forest, the more my everyday worries and frustrations start to fall away.

Under the emerald canopy of leaves, dappled sunlight filters down, transforming the woods into a timeless, almost sacred space. The air feels cleaner here, perfumed with the subtle fragrance of rich soil, wildflowers, and pine. With each breath, I inhale nature's soothing balm.

As I continue deeper into the woods, my senses awaken. The twittering and trilling of birdsong, the skittering of small creatures in the undergrowth, the whisper of a breeze through the branches - these gentle sounds replace the harsh clamor of human activity. Somehow, in the heart of the forest, the important things come back into focus, while the trivial concerns fade into insignificance.

There is a deep peace to be found in the patient presence of trees, standing stalwart and untroubled as they have for ages. When I place my palm against the rough bark of an old oak, I feel connected to something larger than myself, to the unstoppable cycle of the seasons. Beneath my feet, the ancient earth pulses with quiet vitality.

In Japan, there is a concept called "shinrin-yoku", or "forest bathing", which entails spending mindful time in the woods for health benefits. Research has shown that immersing oneself in nature in this way lowers stress, boosts immune function, and promotes mental wellbeing. One need not be a mystic to perceive that there is something deeply healing about retreating to the forest primeval.

More than just a form of exercise or relaxation, walking in the woods can be a spiritual, even transcendent experience. Away from the many screens that monopolize our attention, we have an opportunity to be still, to really observe the miraculous intricacies of the natural world. In this pocket of wild beauty, the essential questions of what it means to be human and how we ought to live arise with newfound clarity.

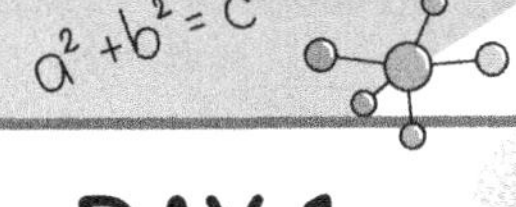

DAY 1
WEEK 7

ELA
READING COMPREHENSION PASSAGE

Here, I remember that for most of human history, it was in places like this that our species found meaning, where we recognized both our smallness and our inextricable connection to creation. As my footsteps fall into a meditative rhythm on the meandering path, I feel myself not so much escaping from the "real world", but rather returning to a truer way of being.

All too soon, my restorative sojourn in the forest must come to an end. But as I turn my steps back towards the trailhead and civilization, I carry the quietude and insights gleaned from my walk in the woods with me. Whenever the burdens of modern existence start to weigh too heavily again, I will know where to go to rediscover what really matters, and find my way back to myself. The woods will be waiting, as they always have, to offer up their profound and timeless teachings.

1. According to the passage, what motivated Henry David Thoreau to go into the woods?

A. To live deliberately
B. To escape his family
C. To find new friends
D. To get more exercise

2. In paragraph 4, the author's description of the sounds of the forest mainly serves to:

A. explain why forests can be dangerous places.
B. highlight how forests differ from urban areas.
C. encourage readers to be louder when in the forest.
D. suggest that forest sounds are harsher than city noises.

3. Which statement best captures how the author feels in the forest?

A. Anxious and unsettled
B. Peaceful and reflective
C. Bored and restless
D. Afraid of the creatures

4. According to the passage, "**forest bathing**" is a practice that involves:

A. taking a bath outdoors in the forest
B. studying the different kinds of trees in forests
C. quickly walking through the forest
D. mindfully spending time in the forest for health benefits

5. The author compares walking in the woods to:

A. playing a competitive sport.
B. taking an exam in school.
C. a spiritual experience.
D. having a party with friends.

6. What effect does placing a palm on an old oak tree have on the author?

A. It makes the author feel connected to something larger.
B. It hurts the author's hand.
C. It makes the author feel powerful and dominant.
D. It scares the author.

7. As used in paragraph 9, the phrase "**profound and timeless teachings**" most nearly means:

A. university-level courses about forests.
B. basic survival skills for the wilderness.
C. important, eternal lessons about life.
D. detailed religious sermons about nature.

8. A central idea of the passage is that:

A. Walking in the woods is a cure for all modern ills.
B. Forests are dangerous places that should be avoided.
C. Spending time in nature can provide mental, physical and spiritual benefits.
D. One must live in the forest full-time to gain any advantages from it.

Let's get some fitness in! Go to page 169 to try some fitness activities.

DAY 2
WEEK 7

ELA
READING COMPREHENSION PASSAGE

Directions: Read the passage below. Then answer the questions.

The Allegory of the Cave

In Book VII of The Republic, Plato presents one of the most famous and influential allegories in Western philosophy - the Allegory of the Cave. Through the character of Socrates, Plato introduces a thought experiment that explores the nature of reality, knowledge, and the philosopher's role in society.

Socrates asks his interlocutor, Glaucon, to imagine a group of prisoners who have been chained in a cave since childhood. They are confined in such a way that they can only look forward at a wall. Behind them, a fire burns, and between the fire and the prisoners is a raised walkway. Along this walkway, people carry various objects, casting shadows on the cave wall that the prisoners perceive as reality. The prisoners, having never seen the actual objects, mistake the shadows for the entirety of existence.

Socrates then supposes that one prisoner is freed and compelled to turn towards the fire. The glare would be disorienting and painful at first, but as his eyes adjusted, he would begin to recognize the objects that had cast the shadows. If he were then dragged out of the cave entirely, he would be blinded by the sun, but gradually, he would acclimate to the world above. He would understand that the sun is the source of life and the true Form of Goodness.

The prisoner, Socrates suggests, would feel compelled to return to the cave and share his revelations with his fellow captives. However, they would likely scorn him, clinging to their familiar understanding of reality. If the enlightened man persisted in trying to liberate them, they might even attempt to kill him.

For Plato, this allegory illuminates fundamental truths about the human condition. The cave represents the world of sensory experience and popular opinion, while the world above symbolizes the realm of abstract, philosophical truth. The average person is content with shadows and illusions, mistaking them for reality. The philosopher, by contrast, strives to free himself from conventional wisdom and perceptive the eternal Forms through reason.

The allegory also speaks to the philosopher's duty to enlighten others, even at great personal risk. Like the freed prisoner, the philosopher who has glimpsed the truth must descend back into the world of shadows to guide others towards wisdom. This is a daunting, even dangerous task, as most people resist challenges to their established beliefs.

ELA

READING COMPREHENSION PASSAGE

On a deeper level, the allegory grapples with the very nature of knowledge and reality. Plato suggests that the world we perceive through our senses is but a shadow of a higher, truer realm. He argues that genuine understanding can only be attained through rigorous philosophical contemplation, not sensory experience. This theory of Forms - the notion of eternal, unchanging absolutes existing beyond the physical world - is central to Platonic thought.

The Allegory of the Cave endures as a seminal work of philosophy, inviting readers to question their assumptions about reality and the limits of their understanding. It challenges us to consider the ways in which our perceptions might be distorted, our beliefs shaped by illusions. At the same time, it charges those who seek truth with the responsibility of sharing their insights, no matter how difficult that may be. In a world all too often fixated on shadows, Plato's allegory remains a powerful call to pursue the light of wisdom.

1. The prisoners in the cave initially mistake the shadows on the wall for:

A. Amusing entertainments
B. Suspicious illusions
C. The entirety of reality
D. Incomplete representations of truth

2. The objects carried along the raised walkway represent:

A. Eternal, unchanging Forms
B. Sensory experiences that most people mistake for reality
C. Useless, trivial items
D. The fundamental building blocks of existence

3. In the allegory, the sun symbolizes:

A. A painful, blinding force
B. A comforting, familiar presence
C. The source of life and the true Form of Goodness
D. A deceptive illusion

4. The prisoner's return to the cave suggests that:

A. The philosopher has a duty to enlighten others, even at personal risk
B. The avg person is eager to have their beliefs challenged
C. Enlightenment is easy to share with others
D. It's best to keep philosophical insights to oneself

5. Plato suggests that genuine understanding comes from:

A. Sensory experiences
B. Rigorous philosophical contemplation
C. Accepting conventional wisdom
D. Studying shadows on cave walls

6. Plato's theory of Forms proposes that:

A. Physical objects are the only true reality
B. Philosophical truth is impossible to attain
C. All knowledge comes from sensory perceptions
D. Eternal, unchanging absolutes exist beyond the physical realm

7. The allegory implies that most people:

A. Are content with illusions and mistake them for reality
B. Actively seek out philosophical truth
C. Spend their lives outside the cave
D. Eagerly accept challenges to their beliefs

8. A central theme of the Allegory of the Cave is:

A. The importance of sensory experience
B. The futility of philosophical contemplation
C. The ease of attaining genuine understanding
D. The philosopher's duty to seek and share truth, however difficult

Let's get some fitness in! Go to page 169 to try some fitness activities.

DAY 3 WEEK 7

MATH LAW OF SINES

OVERVIEW:

Imagine you have an **oblique** triangle (a triangle with no right triangle) with sides of lengths a, b, and c, and the angles opposite these sides are A, B, and C respectively.

The Law of Sines states that the **ratio** of a side length to the sine of its opposite angle is the same for all three sides of the triangle.

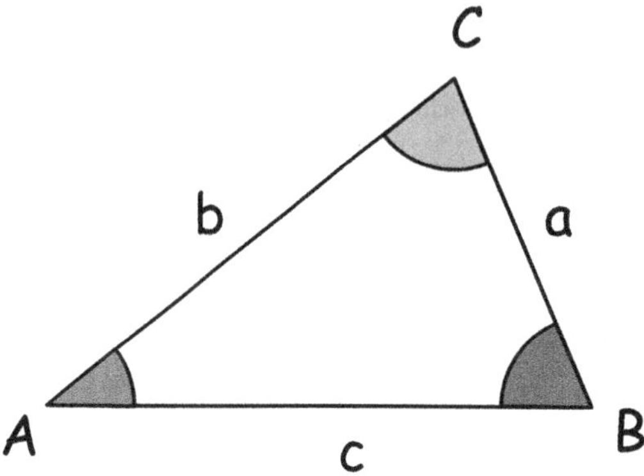

Mathematically:

$$\frac{a}{\sin(A)} = \frac{b}{\sin(B)} = \frac{c}{\sin(C)}$$

Why is this useful?

If you know the lengths of two sides of a triangle and the measure of one of the non-included angles, you can use the Law of Sines to find other angles or sides.

Let's take a look at a few practice examples to understand how to use the Law of Sines formula.

Example 2:

What is $m\angle C$?

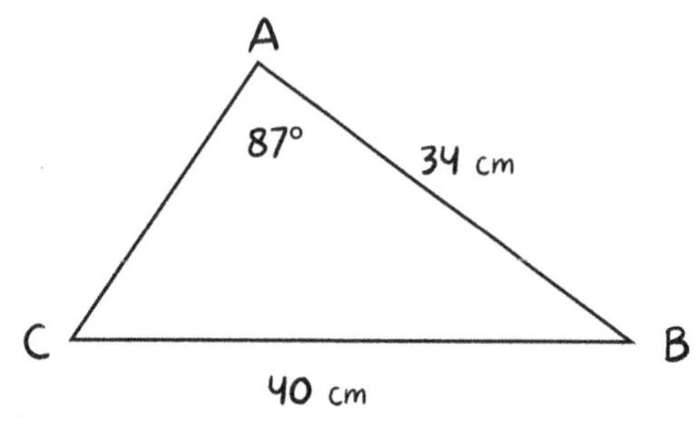

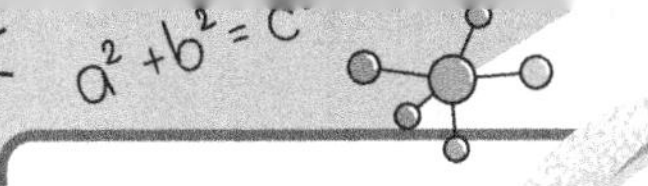

We need to find the angle measure of C. We can use the law of sines to solve this problem.

$\frac{40 \text{ cm}}{\sin(87°)} = \frac{34 \text{ cm}}{\sin(C)}$ ⟵ Cross multiply to get $40 \times \sin(\theta) = 34 \times \sin(87°)$.

You will **need a calculator for these problems.**

$40 \times \sin(C) = 34 \times \sin(87°)$

$40 \times \sin(C) = 34 \times 0.9986$

$40 \times \sin(C) = 33.9534$ ⟵ Now divide both sides by 40

$\sin(C) = 0.8488$ ⟵ How do we solve this? We must use **inverse sin** (arcsin) function on our calculator to get the angle.

arcsin(0.8488)

58.081389672

sin	cos	tan	◉ Deg ○ Rad		7	8	9	+	Back
$\sin^{-1}$	$\cos^{-1}$	$\tan^{-1}$	π	e	4	5	6	–	Ans
x^y	x^3	x^2	e^x	10^x	1	2	3	×	M+
$\sqrt[y]{x}$	$\sqrt[3]{x}$	$\sqrt{x}$	ln	log	0	.	EXP	/	M-
(	)	1/x	%	n!	±	RND	AC	=	MR

$C = 58.1°$ ⟵ We usually always round to the nearest tenths for angles.

m∠C = 58.1°

Example 2:

What is AB?

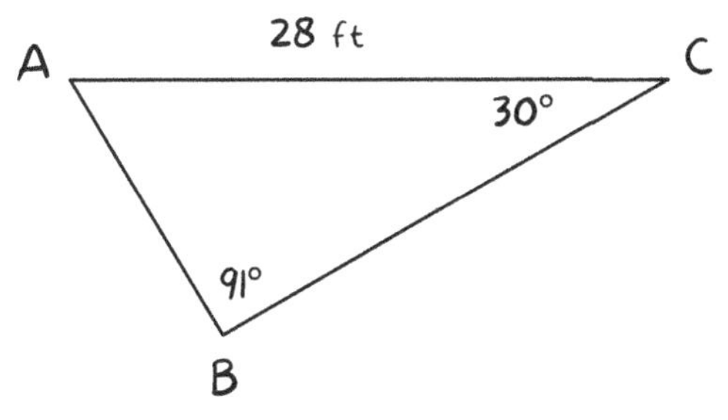

We need to find the length of AB. We can use the law of sines to solve this problem. **Remember**, you can use the Law of Sines if you are provided either

- Two angles and one side (AAS or ASA) OR
- Two sides and an angle opposite one of them (SSA)

DAY 3 WEEK 7

MATH
LAW OF SINES

$\frac{x \text{ ft}}{\sin(30°)} = \frac{28 \text{ ft}}{\sin(91°)}$ ← Cross multiply to get $x(\sin(91°)) = 28(\sin(30°))$

Use your calculator to solve.

$x(\sin(91°)) = 28(\sin(30°))$

$x(0.9998) = 14$ ← Divide both sides by 0.9998

$x = 14.00$ ← We usually round to the nearest hundredths for distance. In this case, the answer is just 14. **Don't forget the units** (feet).

AB = 14 ft

Your turn to practice law of sines!

Directions: Solve these problems using the Law of Sines. You will need to use a calculator. Be sure to show your work on a separate piece of paper.

1. Find $m\angle C$

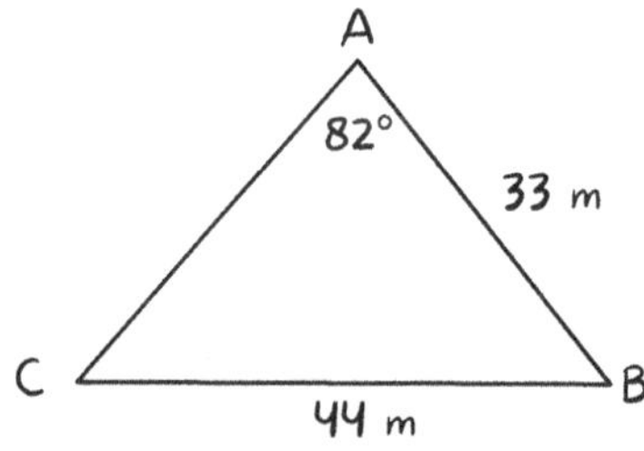

2. Find $m\angle C$

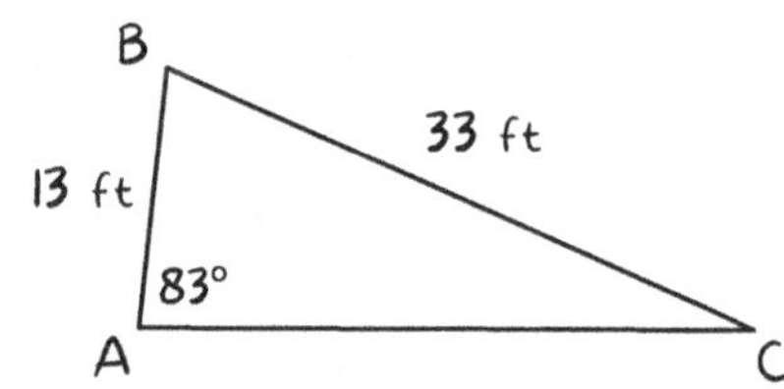

3. Find $m\angle B$

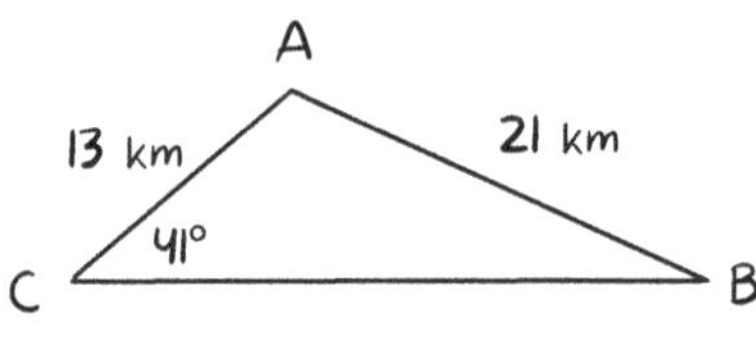

4. Find $m\angle A$

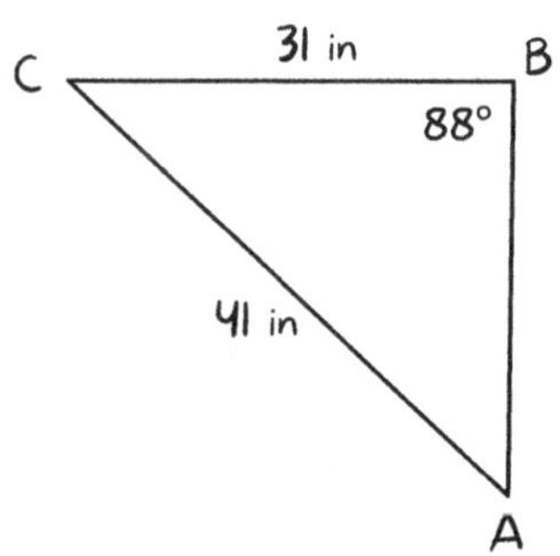

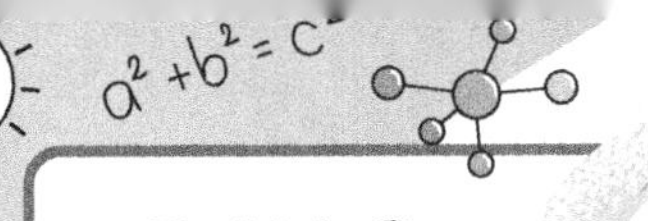

DAY 3
WEEK 7

MATH
LAW OF SINES

5. Find $m\angle B$

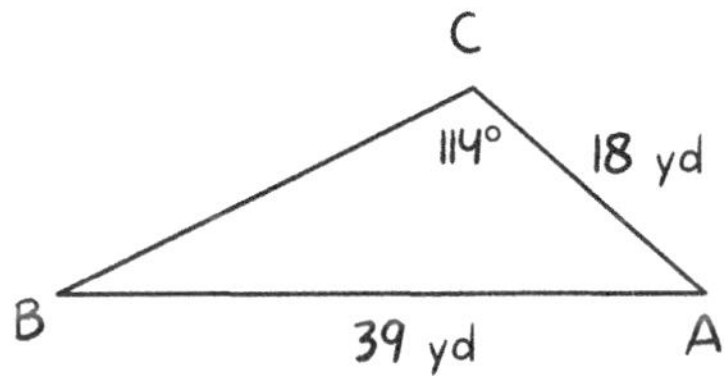

6. Find $m\angle C$

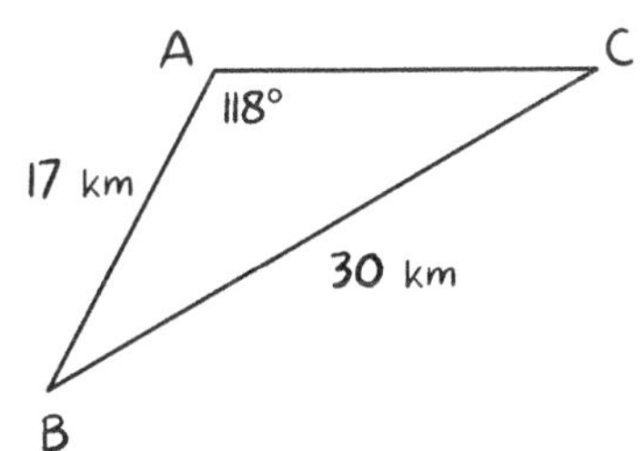

7. Find $m\angle C$

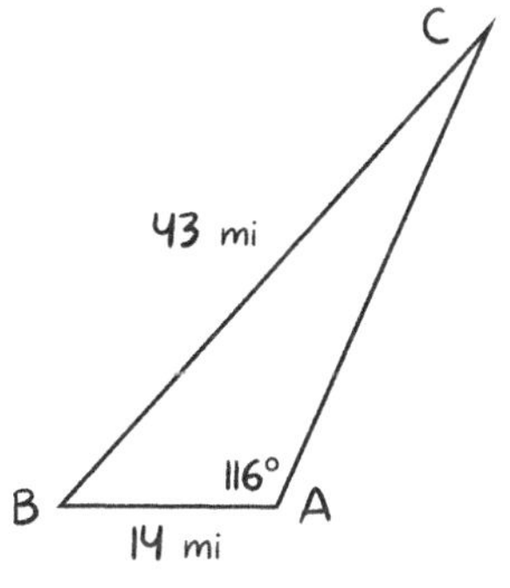

8. Find $m\angle C$

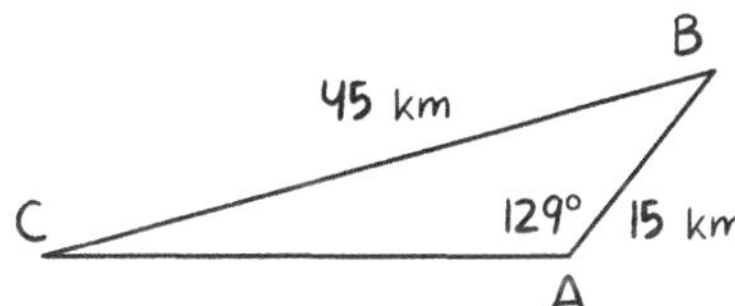

9. Find $m\angle B$

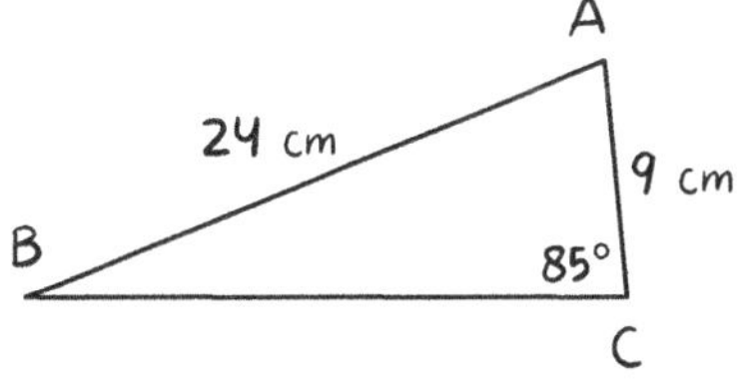

10. Find $m\angle C$

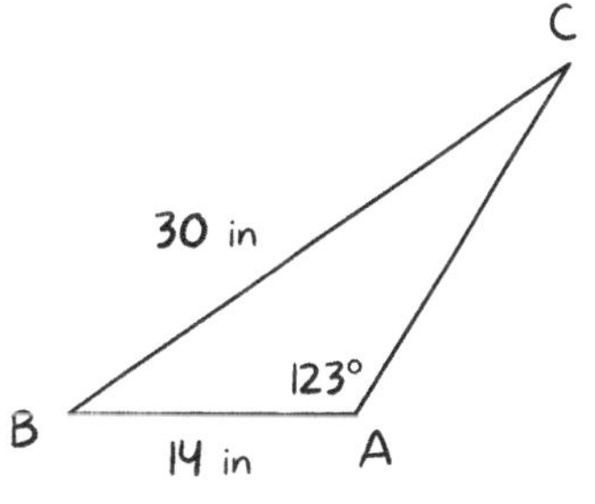

Let's get some fitness in! Go to page 169 to try some fitness activities.

DAY 4
WEEK 7

MATH
LAW OF COSINES

OVERVIEW:

Imagine you have an **oblique** triangle (a triangle with **no** right triangle) with sides of lengths a, b, and c, and the angles opposite these sides are A, B, and C respectively.

The Law of Cosines essentially relates one side of the triangle to the other two sides and the cosine of its included angle.

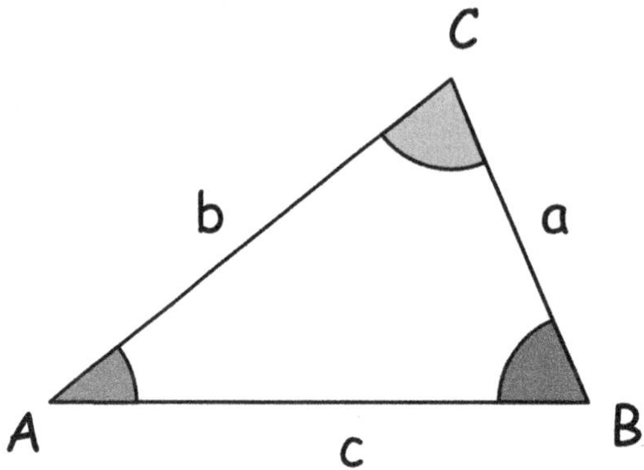

Mathematically:

$$c^2 = a^2 + b^2 - 2ab \cdot \cos(C)$$

This formula can also be written as $a^2 = b^2 + c^2 - 2bc \cos(A)$ **OR** $b^2 = a^2 + c^2 - 2ac \cos(B)$

You just need to remember one of these formulas since all three are the exact same.

You can use the Law of Cosines when you are given the lengths of two sides and the angle between them (SAS) **OR** when you are given the lengths of all three sides (SSS). In these situations, the Law of Sines won't work, as it doesn't provide a feasible ratio to solve.

Let's take a look at a few practice examples to understand how to use the Law of Cosines formula.

Example 1:

What is BC? Round to the nearest tenth.

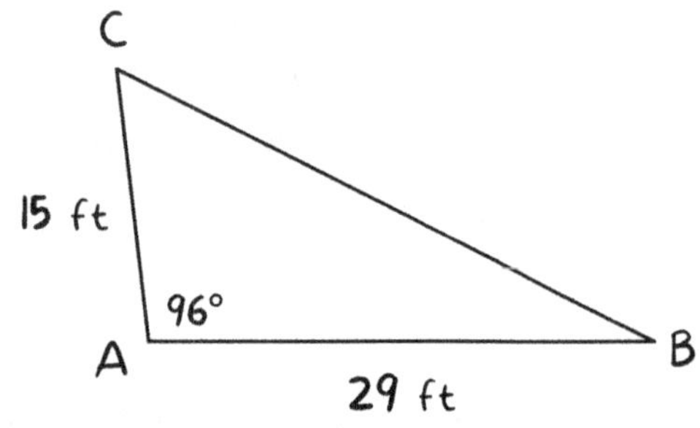

In this example, we are provided with **two sides** and the included angle (SAS).

Let's use our formula to solve.

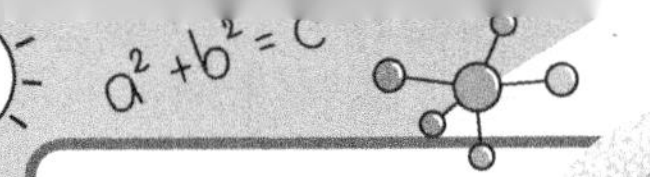

$$c^2 = a^2 + b^2 - 2ab \cdot \cos(C)$$

$$c = \sqrt{a^2 + b^2 - 2ab \cdot \cos(C)}$$

$$c = \sqrt{15^2 + 29^2 - 2(15)(29)(\cos 96^\circ)}$$

$$c = \sqrt{1066 - 870(\cos 96^\circ)}$$

$$c = \sqrt{1066 - (-90.9397)}$$

$$c = \sqrt{1156.9397}$$

$c = 34.01$ ft ⟵ The problem states to round to the nearest tenth so the answer is 34.0 ft, or just 34 ft.

Example 2:

What is BC? Round to the nearest tenth.

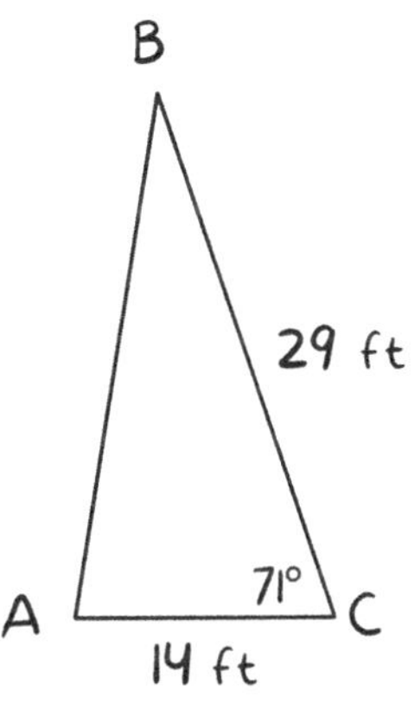

Just like on the example above, we are provided with **two sides** and the included angle (SAS). Let's use our formula to solve.

$$c^2 = a^2 + b^2 - 2ab \cdot \cos(C)$$

$$c = \sqrt{a^2 + b^2 - 2ab \cdot \cos(C)}$$

$$c = \sqrt{14^2 + 29^2 - 2(14)(29)(\cos 71^\circ)}$$

$$c = \sqrt{1037 - 812(\cos 71^\circ)}$$

$$c = \sqrt{1037 - 264.361}$$

$$c = \sqrt{772.639}$$

$c = 27.79$ ft ⟵ The problem states to round to the nearest tenth so the answer is 27.8 ft.

Example 3:

What is m∠C? Round to the nearest tenth.

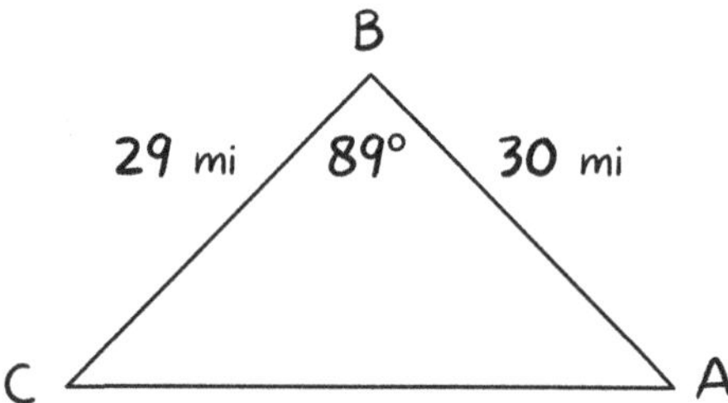

In this example, we are provided with a side, angle and side (SAS). We are asked to find angle C. We **cannot** solve this problem like we solved the previous ones. Notice if we try to use the formula $c^2 = a^2 + b^2 - 2ab\cos(C)$, we have **two unknown** variables. We don't know cos(C) which we are trying to solve for, but more importantly we do not know the value of b. Therefore, we cannot solve this problem until we know the value of b.

So, the first question we ask our self is what is the value or B or length of AC (they are the same thing).

We can use the formula $b^2 = a^2 + c^2 - 2ac\cos(B)$. Plug in the known values.

$$b^2 = 29^2 + 30^2 - 2(29)(30)\cos(89°)$$
$$b^2 = 29^2 + 30^2 - 2(29)(30)\cos(89°)$$
$$b^2 = 1741 - 1740\cos(89°)$$
$$b^2 = 1710.633 \leftarrow \text{Square root both sides}$$
$$b = 41.359 \text{ mi}$$

Let's add in the new information to our diagram. We can round to the nearest hundredths for more accuracy.

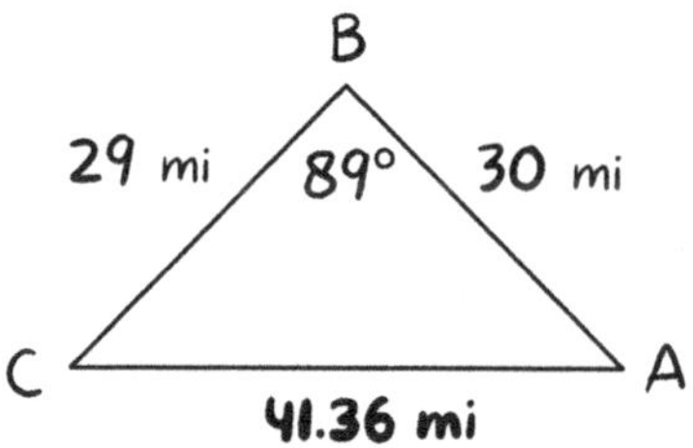

Now we can find m∠C by using the Law of Sines **or** Law of Cosines formula.

DAY 4
WEEK 7

MATH
LAW OF COSINES

If we use the Law of Sines formula, remember $\frac{a}{\sin(A)} = \frac{b}{\sin(B)} = \frac{c}{\sin(C)}$

We can set up the proportional equation $\frac{41.36}{\sin(89°)} = \frac{30}{\sin(C)}$ ← Cross multiply

$41.36 \sin(C) = 30 \sin(89°)$

$41.36 \sin(C) = 29.995$ ← divide both sides by 41.36

$\sin(C) = 0.725$

Use the inverse function (arcsin) on your calculator and we will get $m\angle C = 46.46°$.

We need to round to the nearest tenth as requested in the original problem.

Therefore, $m\angle C = 46.5°$

Your turn to practice law of cosines!

Directions: Solve for the missing sides or angles. Show your work on a separate piece of paper. Round your final answers to the nearest tenth.

1. Find $m\angle B$

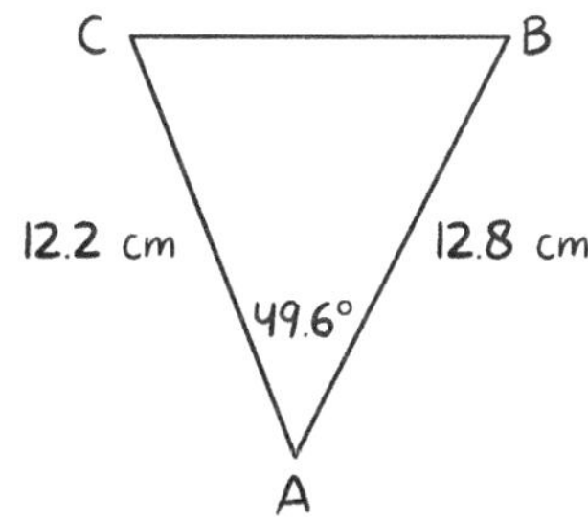

2. Find $m\angle C$

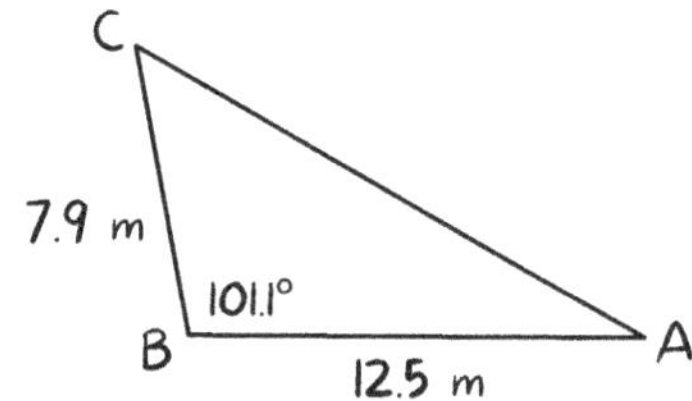

3. Find $m\angle B$

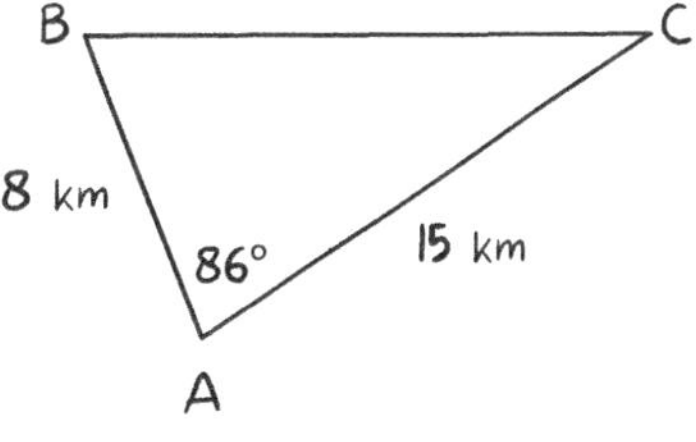

4. Find $m\angle B$

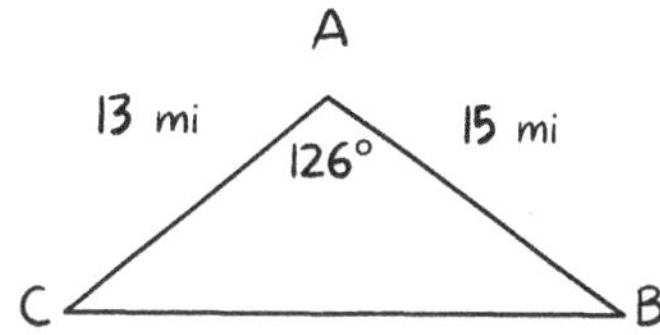

DAY 4 WEEK 7

MATH
LAW OF COSINES

5. Find $m\angle C$

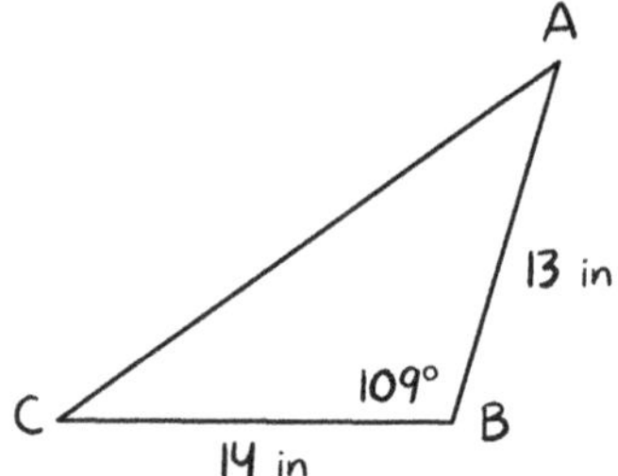

6. Find $m\angle A$

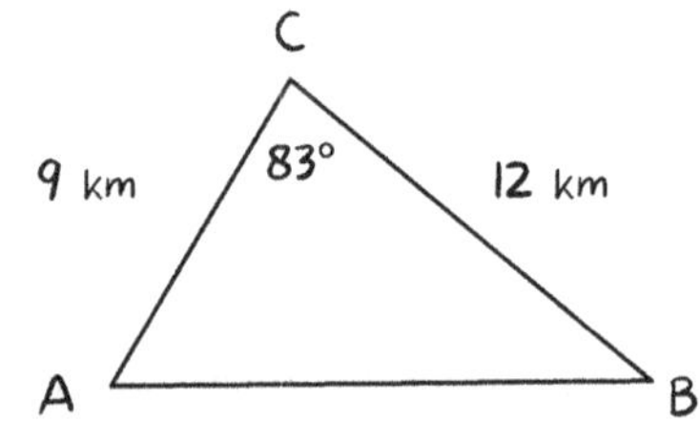

7. Find $m\angle C$

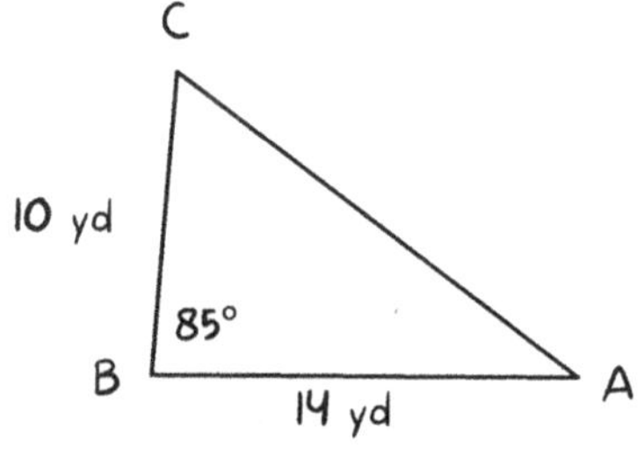

8. Find $m\angle A$

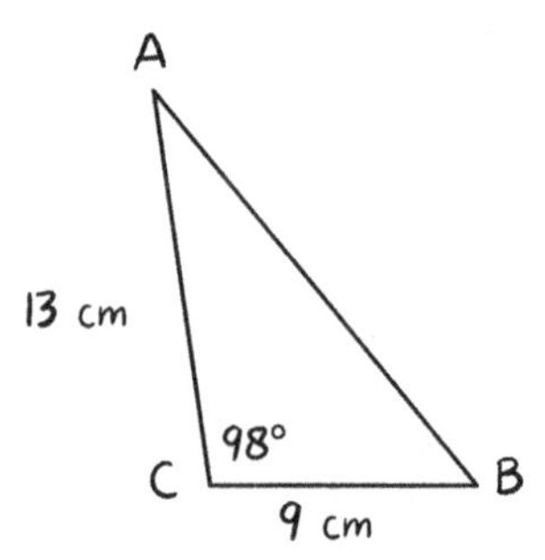

9. Find $m\angle C$

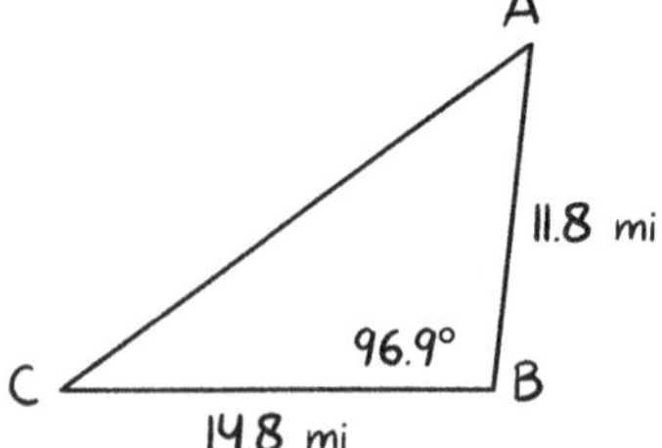

10. Find $m\angle C$

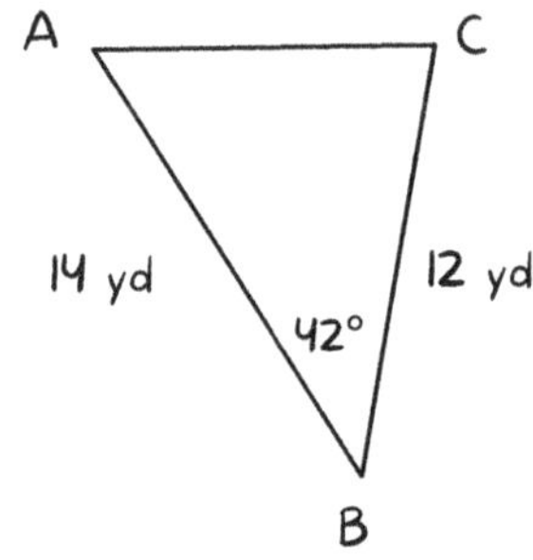

11. Find $m\angle B$

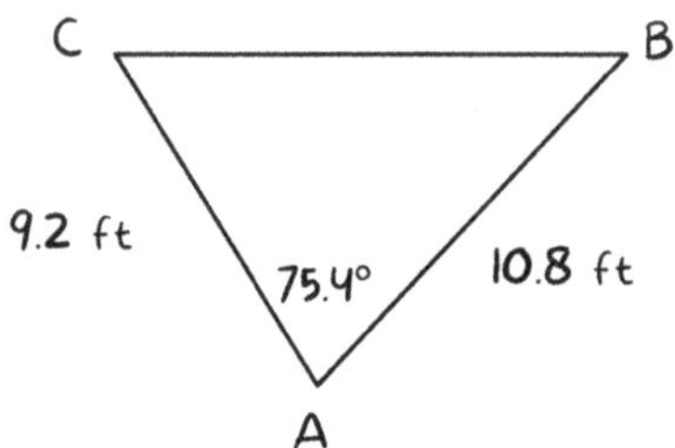

12. Find $m\angle B$

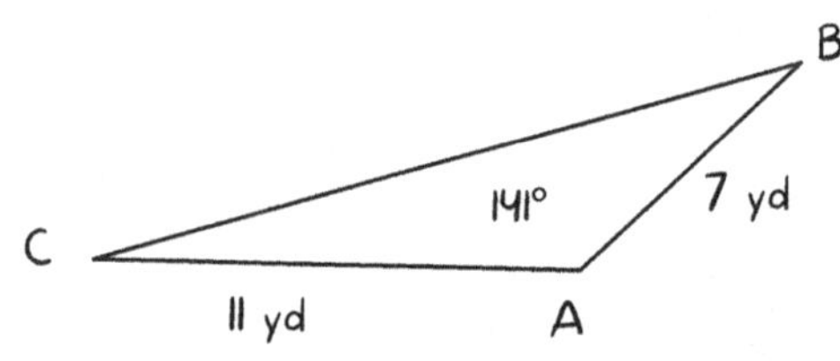

Let's get some fitness in! Go to page 169 to try some fitness activities.

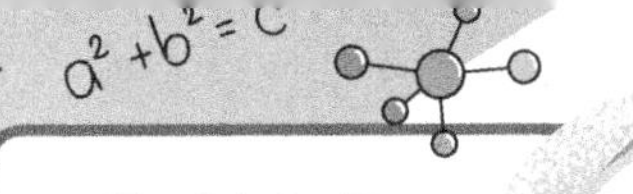

DAY 5 WEEK 7

MATH
ARITHMETIC SEQUENCE

OVERVIEW:

An arithmetic sequence is a sequence of numbers where the difference between the consecutive terms is constant. In other words, we add the same value each time to get the next number in the sequence.

The constant difference between each term is called the "**common difference**" of the sequence. It can be found by subtracting any term in the sequence from the term that follows it.

Example 1:

Consider the sequence: 2, 5, 8, 11, 14, ... In this sequence, each term is 3 more than the previous term. The common difference is 3. In this sequence, what would be the 53rd term?

To solve this, we must use the **Arithmetic Sequence Formula**:

$$[a_n = a_1 + (n - 1) \times d]$$

where,

(a_n) is the n-th term of the sequence.

(a_1) is the first term of the sequence.

(d) is the common difference.

(n) is the position of the term in the sequence.

We are trying to solve for the 53rd term, and we know the other variables. (a_1) is 2, (d) is 3, and (n) is 53. Let's plug that into our equation.

$a_{53} = 2 + (53 - 1) \times 3$

$a_{53} = 2 + (52) \times 3$

$a_{53} = 2 + 156$

$a_{53} = 158$

The 53rd term in this sequence is 158.

DAY 5 WEEK 7

MATH
ARITHMETIC SEQUENCE

Example 2:

In the sequence 26, -174, -374, -574, what is a_{31}?

We need to first find out the common difference between each term which is -200.

Now, let's plug in our known values in the formula.

$a_{31} = 26 + (31 - 1) \times -200$

$a_{31} = 26 + (30) \times -200$

$a_{31} = 26 - 6000$

$a_{31} = -5974$

The 31st term of this sequence is -5974.

Your turn to practice the arithmetic sequence formula!

Notes:

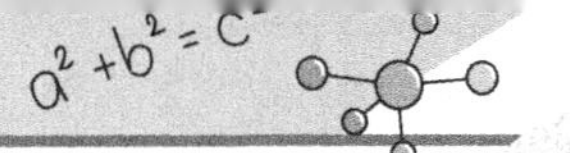

DAY 5
WEEK 7

MATH
ARITHMETIC SEQUENCE

1. $-\frac{13}{9}, \frac{1}{18}, \frac{14}{9}, \frac{55}{18}, \ldots$

Find a_{36}

2. 35, 28, 21, 14, ...

Find a_{36}

3. $-2, -\frac{11}{3}, -\frac{16}{3}, -7, \ldots$

Find a_{23}

4. -13.7, -15, -16.3, -17.6, ...

Find a_{28}

5. 19, 10, 1, -8, ...

Find a_{32}

6. $-\frac{2}{3}, -\frac{7}{3}, -4, -\frac{17}{3}, \ldots$

Find a_{29}

7. -20, -21.7, -23.4, -25.1, ...

Find a_{20}

8. $1, \frac{3}{2}, 2, \frac{5}{2}, \ldots$

Find a_{20}

9. $\frac{6}{5}, \frac{43}{15}, \frac{68}{15}, \frac{31}{5}, \ldots$

Find a_{29}

10. -18, 182, 382, 582, ...

Find a_{40}

11. -11, -21, -31, -41, ...

Find a_{28}

12. $6, \frac{19}{3}, \frac{20}{3}, 7, \ldots$

Find a_{20}

13. -20, -18, -16, -14, ...

Find a_{26}

14. 10, 6, 2, -2, ...

Find a_{35}

15. -8.5, -8.4, -8.3, -8.2, ...

Find a_{22}

Let's get some fitness in! Go to page 169 to try some fitness activities.

Directions: Read each sentence carefully and use context clues to determine the meaning of the underlined word. Choose the best definition for the underlined word from the options provided. Then, write a sentence using the word in a new context, demonstrating your understanding of its meaning and usage. You can use the Internet to look up the exact definition after answering the questions.

1. The journalist's **perspicacious** analysis of the political situation revealed hidden connections and potential consequences that other reporters had overlooked.

   A. shallow
   B. biased
   C. insightful
   D. confusing

   Sentence:

2. The company's **byzantine** organizational structure made it difficult for employees to navigate the chain of command and get clear answers to their questions.

   A. straightforward
   B. efficient
   C. lazy
   D. convoluted

   Sentence:

3. The **peripatetic** lifestyle of a traveling salesman suited him well, as he found the constant change of scenery and the opportunity to meet new people **invigorating**.

A. sedentary; draining
B. nomadic; energizing
C. comfortable; boring
D. stressful; intimidating

Sentence:

4. The company's **ignominious** collapse was precipitated by a series of scandals that exposed widespread corruption and financial mismanagement at the highest levels of leadership.

A. dishonorable
B. gradual
C. celebrated
D. unexpected

Sentence:

5. The **inveterate** gambler found it nearly impossible to resist the allure of the casino, often spending hours at the tables in pursuit of an elusive big win, despite the mounting financial and personal costs.

A. occasional
B. skilled
C. reluctant
D. habitual

Sentence:

Let's get some fitness in! Go to page 169 to try some fitness activities.

Directions: Work through these questions carefully.

Passage 1:

In the midst of the Enlightenment, a period characterized by the celebration of reason and the pursuit of knowledge, there emerged a curious figure whose ideas would challenge the very foundations of conventional thought. Born in 1788, Arthur Schopenhauer was a German philosopher who, in contrast to his contemporaries, posited a world view that was deeply pessimistic and marked by a profound sense of the inherent suffering of human existence. Schopenhauer's philosophy, heavily influenced by Eastern thought, particularly Buddhism, stood in stark contrast to the prevailing Enlightenment ideals of progress, optimism, and the power of rational thought to unlock the mysteries of the universe and improve the human condition.

1. The passage suggests that Schopenhauer's ideas were:

   A. Consistent with Enlightenment principles
   B. Influenced by Western philosophy
   C. Optimistic about the human condition
   D. Contrary to prevailing thought of the time

2. Which of the following best describes the tone of the passage?

   A. Laudatory
   B. Objective
   C. Derisive
   D. Nostalgic

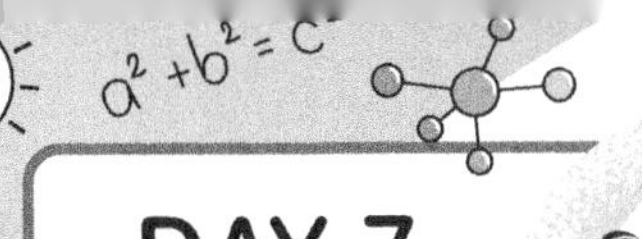

Passage 2:

The concept of "manifest destiny," which asserted that the United States was destined to expand across the North American continent, was a potent force in shaping American foreign policy and justifying territorial expansion in the 19th century. While proponents of manifest destiny portrayed it as a benevolent mission to spread democracy and civilization, critics argue that it was little more than a thinly veiled justification for imperialism, racism, and the displacement of indigenous peoples. The legacy of manifest destiny continues to be a subject of debate among historians, with some seeing it as a defining aspect of American national identity and others viewing it as a shameful chapter in the nation's history that must be reckoned with.

3. The passage suggests that manifest destiny was:

A. Universally accepted by Americans
B. A justification for territorial expansion
C. A benevolent mission to spread democracy
D. Unrelated to racism or displacement of indigenous peoples

4. The author's perspective on the legacy of manifest destiny can best be described as:

A. Unequivocally positive
B. Dismissive of its impact
C. Acknowledging its complexity and controversies
D. Uninterested in its historical significance

5. In the context of the passage, the word "reckoned with" most nearly means:

A. Embraced
B. Ignored
C. Challenged
D. Confronted and addressed

Let's get some fitness in! Go to page 169 to try some fitness activities.

WEEK 8
GRADE 10-11
In this last week, you will practice writing an essay after conducting thorough research. For math, you will learn about geometric sequences and probability. Congrats on making it to the final week of the workbook!
ARGOPREP

DAY 1
WEEK 8

ELA
READING COMPREHENSION PASSAGE

Directions: Read the passage below. Then answer the questions.

The Paradoxical Plight of Genetic Enhancement

The rapid advancements in genetic engineering have given rise to a perplexing ethical quandary: the potential for enhancing human traits and abilities through genetic modification. Proponents argue that genetic enhancement could lead to a new era of human progress, eliminating diseases, increasing intelligence and physical capabilities, and even extending lifespans. Detractors, however, warn of the dire consequences that could ensue, from exacerbating social inequalities to undermining the very essence of what it means to be human.

At the heart of this debate lies a fundamental question: What is the nature of human flourishing, and how do we best achieve it? The allure of genetic enhancement stems from the belief that by optimizing our biological traits, we can transcend our limitations and achieve greater fulfillment. If we can boost our cognitive abilities, eradicate vulnerabilities to disease, and prolong our vitality, the argument goes, we will be better equipped to pursue our goals, realize our potential, and lead rich, meaningful lives.

However, this sanguine view overlooks the complex web of factors that contribute to human well-being. While biological traits undoubtedly play a role, they are far from the only determinants of a life well-lived. Psychological resilience, emotional intelligence, strong social bonds, a sense of purpose—these are just a few of the many elements that shape our overall flourishing. By fixating on genetic enhancement as a panacea, we risk neglecting the cultivation of these vital nonbiological dimensions.

Moreover, the pursuit of genetic perfection could have unintended consequences that undermine the very well-being we seek to promote. The knowledge that one's abilities are the result of engineered enhancements rather than earned through effort and perseverance could erode self-esteem and diminish the sense of authorship over one's accomplishments. Furthermore, as genetic enhancements become more prevalent, those who lack access to these technologies may face even greater disadvantages, exacerbating societal disparities and creating a new form of discrimination based on biological pedigree.

Perhaps most troubling is the potential for genetic enhancement to homogenize humanity and narrow our conception of what it means to flourish. If we all strive for the same optimized traits, we risk sacrificing the rich diversity of perspectives, talents, and ways of being that characterize the human experience. In a world of genetically enhanced individuals, will there still be room for the unique quirks, flaws, and idiosyncrasies that make us each distinctly human?

As we grapple with these thorny questions, we must approach the prospect of genetic enhancement with great caution and humility. While the potential benefits are tantalizing, the risks and ethical pitfalls are grave. We must resist the temptation to view our biology as the sole determinant of our flourishing and recognize the multifaceted nature of human well-being.

ELA
READING COMPREHENSION PASSAGE

1. Which of the following does the author NOT present as a potential argument in favor of genetic enhancement?

A. Genetic enhancement could help eliminate diseases and extend human lifespans.
B. Optimizing biological traits could allow individuals to achieve greater personal fulfillment.
C. Genetic enhancement would promote greater social equality and reduce discrimination.
D. Boosting cognitive abilities through genetic modification could enhance individuals' pursuit of their goals.

2. As used in line 17, the word "panacea" most nearly means:

A. a dangerous practice
B. a beneficial supplement
C. a complex process
D. a cure-all solution

3. The author implies that a potential consequence of widespread genetic enhancement could be:

A. a reduction in societal disparities and discrimination
B. an increased sense of self-esteem and accomplishment
C. a loss of diversity in human traits and perspectives
D. a greater emphasis on cultivating psychological well-being

4. Analyze the ethical implications of genetic enhancement as discussed in the passage. Do you believe the potential benefits justify the ethical concerns? Support your answer with specific details from the text.

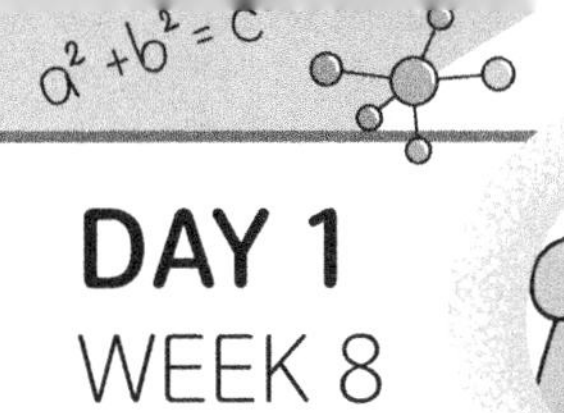

ELA

READING COMPREHENSION PASSAGE

Let's get some fitness in! Go to page 169 to try some fitness activities.

DAY 2
WEEK 8

ELA
WRITING AN ESSAY

Essay Topic:

Examine the long-term effects of the COVID-19 pandemic on mental health worldwide, focusing on specific demographics such as teenagers or healthcare workers.

Instructions:

1. Research: Look into medical studies, surveys, and health reports published since the outbreak of the pandemic.
2. Thesis Development: Develop a thesis on the significant mental health trends emerging from the pandemic within your chosen demographic.
3. Evidence-Based Support: Present statistical data and personal accounts or interviews if possible.
4. Solutions and Support Systems: Analyze the effectiveness of mental health support systems introduced during and after the pandemic.
5. Conclusion: Discuss the potential lasting impacts on the demographic and mental health services.
6. Citations: Use MLA or APA format to cite all your research sources accurately.

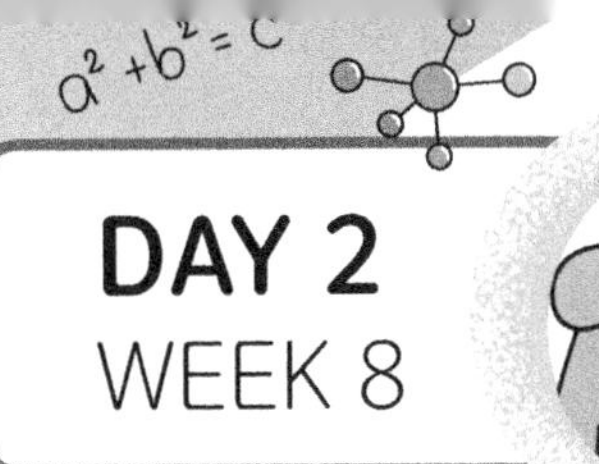

DAY 2
WEEK 8

ELA

WRITING AN ESSAY

ELA
WRITING AN ESSAY

Let's get some fitness in! Go to page 169 to try some fitness activities.

FITNESS TIME

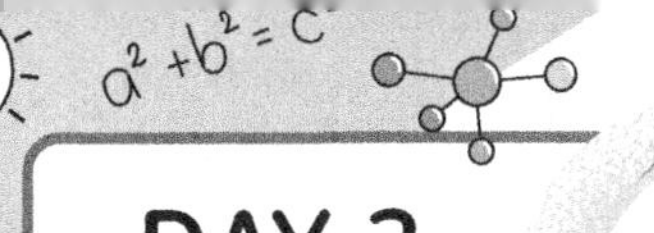

DAY 3
WEEK 8

MATH
GEOMETRIC SEQUENCES

OVERVIEW:

A geometric sequence is a series of numbers where each term after the first is found by multiplying the previous one by a fixed, non-zero number called the common ratio. The formula for the nth term of a geometric sequence can be expressed as:

$$a_n = a \times r^{(n-1)}$$

where,

(a) is the first term of the sequence,

(r) is the common ratio,

(n) is the term number

(a_n) is the nth term of the sequence

As an example, a sequence starting with 2 and having a common ratio of 3 would have the following sequence: 2, 6, 18, 54, 162, ...

What if we needed to termine the 10th term?

We can use our geometric formula and plug in our known values.

a is the first term which is 2, r is the common ratio which is 3, n is 10.

$a_{10} = 2 \times 3^{(10-1)}$

$a_{10} = 2 \times 3^{(9)}$

$a_{10} = 2 \times 19683$

$a_{10} = 39{,}366$

The 10th term in this sequence would be 39,366.

Your turn to practice the geometric sequence formula!

DAY 3 WEEK 8

MATH
GEOMETRIC SEQUENCES

1. 3, 12, 48, 192, ...

Find a_{10}

2. 1, -4, 16, -64, ...

Find a_{10}

3. -2, -4, -8, -16, ...

Find a_{10}

4. 1, 3, 9, 27, ...

Find a_{9}

5. 1, -5, 25, -125, ...

Find a_{9}

6. 1, -2, 4, -8, ...

Find a_{12}

7. 2, -4, 8, -16, ...

Find a_{12}

8. -3, -9, -27, -81, ...

Find a_{9}

9. -4, -8, -16, -32, ...

Find a_{11}

10. 3, -12, 48, -192, ...

Find a_{9}

11. 1, 2, 4, 8, ...

Find a_{10}

12. -3, 9, -27, 81, ...

Find a_{11}

13. 4, -12, 36, -108, ...

Find a_{9}

14. -2, -8, -32, -128, ...

Find a_{10}

15. 4, 8, 16, 32, ...

Find a_{11}

Let's get some fitness in! Go to page 169 to try some fitness activities.

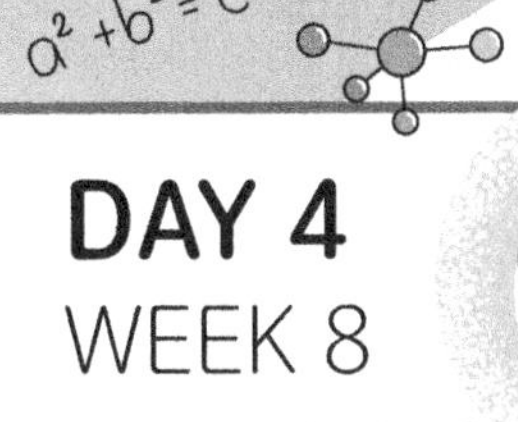

DAY 4
WEEK 8

MATH
PROBABILITY

OVERVIEW:

Probability is a measure of the likelihood that an event will occur, expressed as a number between 0 and 1. Probability is defined as the ratio of the number of favorable outcomes to the number of possible outcomes when all outcomes are equally likely. The formula for probability is:

$$P(E) = \frac{\text{Number of favorable outcomes}}{\text{Total number of possible outcomes}}$$

Example:

Suppose you toss a fair coin. There are two possible outcomes: heads or tails. The probability of getting heads is $P(Heads) = \frac{1}{2}$ since there is one favorable outcome (heads) and two possible outcomes (heads, tails).

Here are some common knowledge you should already be familiar with.

1. Coin flip:
- A fair coin has two possible outcomes: heads or tails.
- The total number of possible outcomes for a single coin flip is 2.

2. Fair six-sided die:
- A fair six-sided die has six possible outcomes: 1, 2, 3, 4, 5, or 6.
- The total number of possible outcomes for a single roll of a fair six-sided die is 6.

3. Deck of cards:
- A standard deck of cards consists of 52 cards.
- There are four suits: hearts, diamonds, clubs, and spades.
- Each suit has 13 cards: Ace, 2, 3, 4, 5, 6, 7, 8, 9, 10, Jack, Queen, and King.
- The total number of possible outcomes when drawing a single card from a well-shuffled deck is 52.

4. Spinners:
- A spinner is a circular device divided into several equal-sized sections, each representing a possible outcome.
- For example, a spinner with four equal sections labeled A, B, C, and D has four possible outcomes.
- The total number of possible outcomes for a single spin is 4.

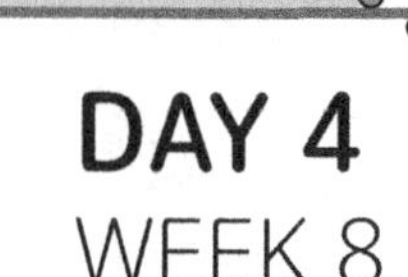

5. Bag of marbles:
   - A bag containing marbles of different colors can be used to demonstrate probability concepts.
   - For example, a bag with 3 red marbles, 2 blue marbles, and 5 green marbles has a total of 10 marbles.
   - The total number of possible outcomes when drawing a single marble from the bag is 10.

Great, let's review Compound Events!

Compound Events:

When dealing with more than one event, the calculation of probabilities can involve:

- Independent Events: Events where the outcome of one does not affect the outcome of another.
- Dependent Events: Events where the outcome of one event affects the outcome of another.

Example of Independent Events:

Tossing a coin and rolling a die. The outcome of the coin toss does **not** affect the die roll.

What is the probability that you will land on heads and roll on a 4?

$$P(\text{Heads and } 4) = P(\text{Heads}) \times P(4) = \frac{1}{2} \times \frac{1}{6} = \frac{1}{12}$$

The probability is $\frac{1}{12}$.

Example of Dependent Events:

Drawing two cards from a deck without replacement. The outcome of the first draw affects the probabilities in the second draw.

What is the probability you will draw an Ace on the first draw and King on the second draw?

$$P(\text{Drawning an Ace and then a King}) = P(\text{Ace}) \times P(\text{King} \mid \text{Ace}) = \frac{4}{52} \times \frac{4}{51} = \frac{16}{2652} = \frac{4}{663}$$

Your turn to practice probability word problems!

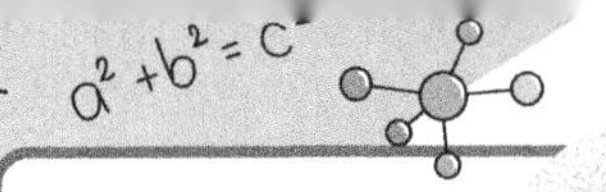

DAY 4
WEEK 8

MATH
PROBABILITY

1. Maya and Lucas are playing a dice game where they roll two six-sided dice. The winner is the person who rolls a pair of dice that sums to exactly 7. If they play this game once, what is the probability that Maya wins on her first try?

2. Sam shuffles a standard deck of 52 cards thoroughly and draws a card from the top. What is the probability that the card drawn is either a red card or a face card (jack, queen, king)?

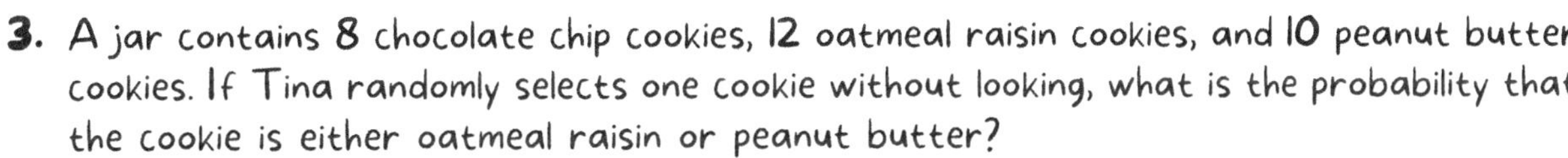

3. A jar contains 8 chocolate chip cookies, 12 oatmeal raisin cookies, and 10 peanut butter cookies. If Tina randomly selects one cookie without looking, what is the probability that the cookie is either oatmeal raisin or peanut butter?

Let's get some fitness in! Go to page 169 to try some fitness activities.

DAY 5 WEEK 8

MATH PROBABILITY

Work through these probability word problems carefully.

1. The school is conducting a raffle. There are **500** tickets sold, and Jamie bought **10** of them. The school is drawing **3** tickets to determine the winners of **3** different prizes. What is the probability that **at least** one of Jamie's tickets is drawn? Assume that once a ticket is drawn, it is not put back into the pool.

   Hint: (Calculate the probability of Jamie not winning with any ticket and subtract this from 1 to find the probability of winning at least once.)

2. A school library categorizes its books into three sections: fiction, non-fiction, and reference. There are **120** fiction books, **80** non-fiction books, and **40** reference books. If a student randomly selects one book, what is the probability that it is either a fiction or a non-fiction book?

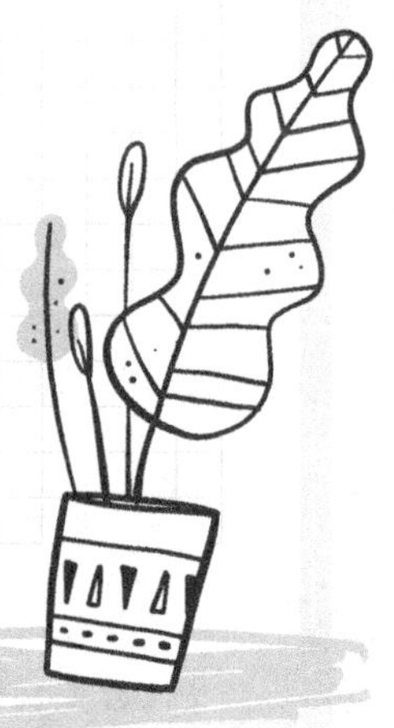

MATH
PROBABILITY

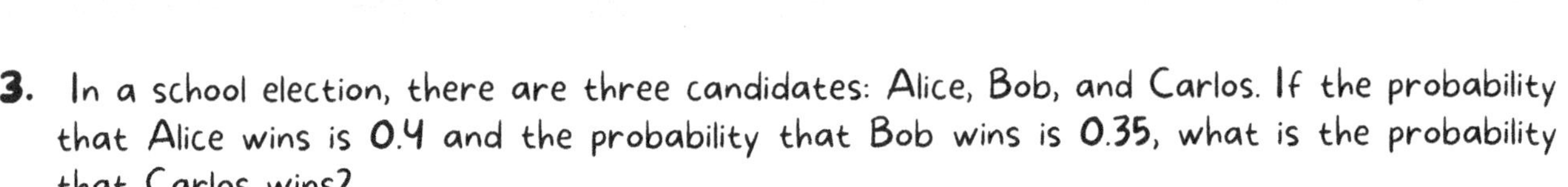

3. In a school election, there are three candidates: Alice, Bob, and Carlos. If the probability that Alice wins is 0.4 and the probability that Bob wins is 0.35, what is the probability that Carlos wins?

4. When rolling two six-sided dice, what is the probability that the sum of the numbers rolled is either 7 or 11?

5. In a survey of 200 students, 80 like apples, 70 like oranges, and 30 like both. What is the probability that a student chosen at random likes either an apple or an orange?

Let's get some fitness in! Go to page 169 to try some fitness activities.

DAY 6
WEEK 8

ELA
VOCABULARY

Directions: Read each sentence carefully and use context clues to determine the meaning of the underlined word. Choose the best definition for the underlined word from the options provided. For Questions 1-3, write a sentence using the word in a new context, demonstrating your understanding of its meaning and usage. You can use the Internet to look up the exact definition after answering the questions.

1. The **recondite** symbology employed by the ancient civilization confounded archaeologists for decades until a breakthrough discovery provided the key to deciphering the enigmatic language.

A. transparent **B.** abstruse **C.** conventional **D.** simplistic

Sentence:

....

....

....

2. The **assiduous** student spent countless hours poring over textbooks, meticulously taking notes, and diligently completing assignments, all in the pursuit of academic excellence.

A. lazy
B. diligent
C. indifferent
D. sporadic

Sentence:

....

....

....

3. The **quixotic** quest to find a cure for the rare disease consumed the researcher, who remained undeterred by the countless setbacks and the seemingly insurmountable odds.

A. realistic
B. straightforward
C. unimportant
D. impractical

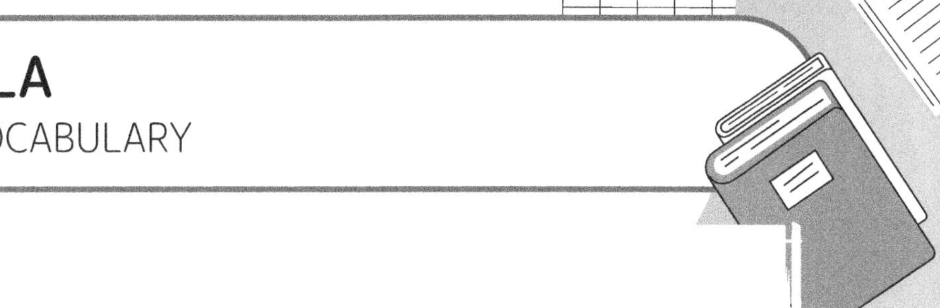

Sentence:

4. The **quiescent** volcano, which had remained dormant for centuries, suddenly **erupted** with a **cataclysmic** force, spewing ash and lava into the sky and transforming the once-peaceful landscape into a scene of **primordial** chaos.

A. active; subsided; minor; modern

B. extinct; smoldered; gradual; futuristic

C. dormant; exploded; devastating; ancient

D. unstable; fizzled; anticlimactic; contemporary

Sentence:

5. The **punctilious** accountant meticulously reviewed every detail of the company's financial records, ensuring that each transaction was **scrupulously** documented and that not a single penny was unaccounted for, much to the **chagrin** of his less detail-oriented colleagues.

A. meticulous; carefully; annoyance

B. careless; haphazardly; delight

C. efficient; sporadically; indifference

D. disorganized; recklessly; relief

Let's get some fitness in! Go to page 169 to try some fitness activities.

DAY 7 WEEK 8

SAT ELA PREP

Directions: Work through these questions carefully.

1. In the sentence "The scientist's groundbreaking research, which was conducted over several years, revolutionized the field of genetics," what is the function of the phrase "which was conducted over several years"?

   A. Independent clause
   B. Subordinate clause
   C. Prepositional phrase
   D. Appositive phrase

2. Select the sentence that contains a dangling modifier:

   A. Having finished the experiment, the results were analyzed by the team.
   B. The results were analyzed by the team, who had finished the experiment.
   C. The team analyzed the results, having finished the experiment.
   D. After finishing the experiment, the team analyzed the results.

3. Choose the word that best completes the sentence: The politician's ___________ speech left the audience feeling inspired and motivated.

   A. pedantic
   B. mundane
   C. stirring
   D. insipid

4. Select the sentence that contains a correct use of a semicolon:

   A. The band played their hit song; the crowd erupted in applause.
   B. The chef prepared a delicious meal; consisting of three courses.
   C. The hiker reached the summit; exhausted but exhilarated.
   D. The student studied diligently; however, she still felt unprepared for the exam.

5. Choose the sentence that contains an error in parallel structure:

   A. The teacher encouraged her students to read widely, think critically, and write persuasively.
   B. The restaurant offers a variety of dishes, including pasta, seafood, and they also have vegetarian options.
   C. The athlete trained rigorously, ate a balanced diet, and maintained a positive attitude.
   D. The company seeks employees who are creative, dedicated, and possess strong problem-solving skills.

Let's get some fitness in! Go to page 169 to try some fitness activities.

FITNESS TIME

FITNESS
TIME
GO!
ARGOPREP

FITNESS TIME

exercises complex one

1 - **Abs:** 3 times

2 - **Lunges:** 2 times to each leg.

Note: Use your body weight or books as weight to do leg lunges.

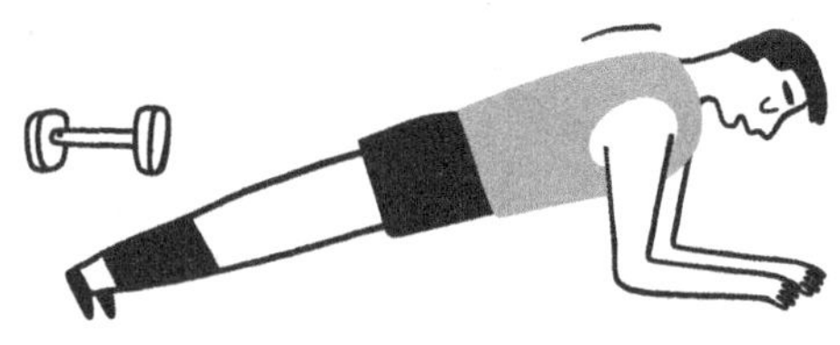

3 - **Plank:** 6 sec.

4 - **Run:** 50 m

Note: Run 25 meters to one side and 25 meters back to the starting position.

> Please be aware of your environment and be safe at all times.
> If you cannot do an exercise, just try your best.

exercises complex two

1 - **High Plank:** 6 sec.

3 - **Waist Hooping:** 10 times.

Note: if you do not have a hoop, pretend you have an imaginary hoop and rotate your hips 10 times.

2 - **Chair:** 10 sec.

Note: sit on an imaginary chair while keeping your back straight.

4 - **Abs:** 10 times

FITNESS TIME

exercises complex three

2 - **Bend Down:** 10 sec.

3 - **Chair:** 10 sec.

1 - **Down Dog:** 10 sec.

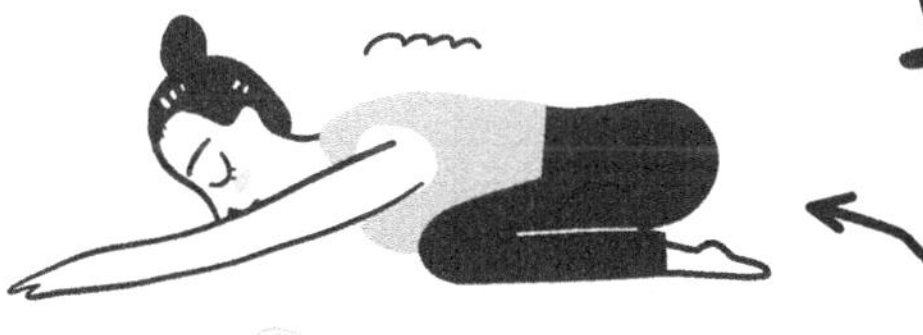

5 - **Shavasana:** as long as you can.
Note: think of happy moments and relax your mind.

4 - **Child Pose:** 20 sec.

Please be aware of your environment and be safe at all times.
If you cannot do an exercise, just try your best.

exercises complex four

2 - **Lunges:** 3 times to each leg.

Note: Use your body weight or books as weight to do leg lunges.

1 - **Bend forward:** 10 times.

Note: try to touch your feet. Make sure to keep your back straight, and if needed, you can bend your knees.

3 - **Plank:** 6 sec.

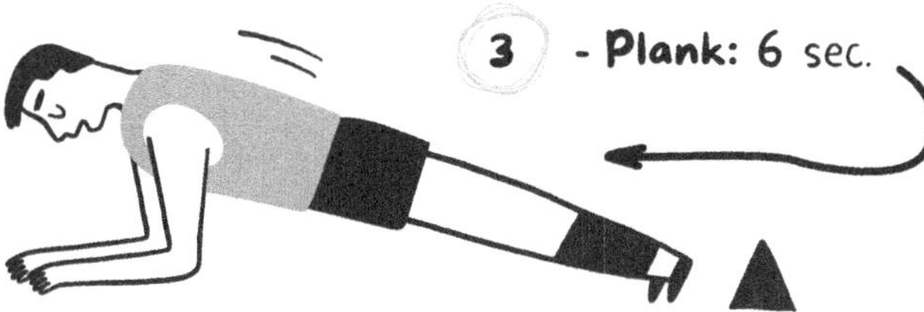

4 - **Abs:** 10 times

Answer Sheets

To see the answer key to the entire workbook, you can easily download the answer key from our website!

*Due to the high request from parents and teachers, we have removed the answer key from the workbook so you do not need to rip out the answer key while students work on the workbook.

Go to **argoprep.com/summer11**

OR scan the QR Code:

Place your mouse over the workbook you have, and you will see the "Download Answers" button.

For detailed video instructions on how to access the "Answer Sheets," please scan this QR code.

Download Answers

Kids Summer Academy by ArgoPrep: Grade 8-9

Kids Summer Academy by ArgoPrep: Grade 5-6

Kids Summer Academy by ArgoPrep: Grade 4-5

Kids Summer Academy by ArgoPrep: Grade 6-7

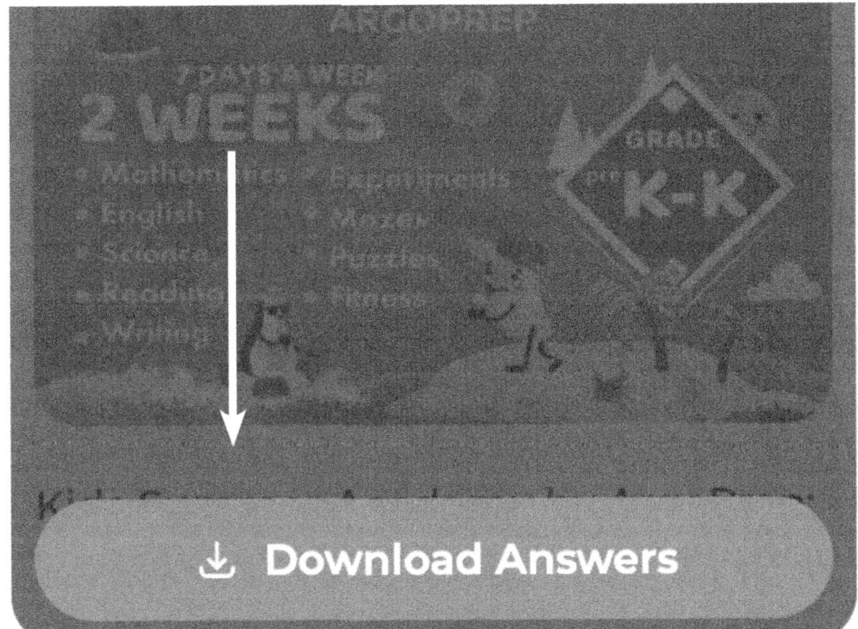
Download Answers

7 DAYS A W
2 WEEK
Mathematics
English
Science
Reading
Writing
Kids Summer
Grade 8-9

Made in United States
Orlando, FL
30 April 2025

60892044R00096